AF729479

Earthen Construction Technology

# Earthen Construction Technology

Proceedings of the XVIII UISPP World Congress
(4-9 June 2018, Paris, France)
Volume 11

Session IV-5

edited by

Annick Daneels and Maria Torras Freixa

Archaeopress Archaeology

Archaeopress Publishing Ltd
Summertown Pavilion
18-24 Middle Way
Summertown
Oxford OX2 7LG

www.archaeopress.com

ISBN 978-1-78969-723-0
ISBN 978-1-78969-724-7 (e-Pdf)



Cover: Greater Kyz Kala Fortress in Ancient Merv (Turkmenistan) (Photograph by John Pavelka, 2018, CC-BY-2.0)

This book is available direct from Archaeopress or from our website www.archaeopress.com

UISPP PROCEEDINGS SERIES VOLUME 11 - Earthen Construction Technology

UISPP XVIII CONGRESS, PARIS, JUNE 2018

(4-9 June 2018, Paris)

Session IV-5

VOLUME EDITORS: Annick Daneels and Maria Torras Freixa

SERIES EDITOR: The Board of UISPP

SERIES PROPERTY: UISPP – International Union of Prehistoric and Protohistoric Sciences


KEY-WORDS IN THIS VOLUME: Architectural traditions, Mudbrick, Cob, Wattle-and-daub, Archaeometry

UISPP PROCEEDINGS SERIES is a printed on demand and an open access publication, edited by UISPP through Archaeopress



2018 Paris
4-9 juin
uispp
Union Internationale des Sciences Préhistoriques et Protohistoriques
XVIIIe Congrès

# FOREWORD TO THE XVIII UISPP CONGRESS PROCEEDINGS

UISPP has a long history, originating in 1865 in the International Congress of Prehistoric Anthropology and Archaeology (CIAAP). This organisation ran until 1931 when UISPP was founded in Bern. In 1955, UISPP became a member of the International Council of Philosophy and Human Sciences, a non-governmental organisation within UNESCO.

UISPP has a structure of more than thirty scientific commissions which form a very representative network of worldwide specialists in prehistory and protohistory. The commissions cover all archaeological specialisms: historiography; archaeological methods and theory; material culture by period (Palaeolithic, Neolithic, Bronze Age, Iron Age) and by continents (Europe, Asia, Africa, Pacific, America); palaeoenvironment and palaeoclimatology; archaeology in specific environments (mountain, desert, steppe, tropical); archaeometry; art and culture; technology and economy; biological anthropology; funerary archaeology; archaeology and society.

The UISPP XVIII World Congress of 2018 was hosted in Paris by the University Paris 1 Panthéon-Sorbonne with the strong support of all French institutions related to archaeology. It featured 122 sessions, and over 1800 papers were delivered by scientists from almost 60 countries and from all continents.

The proceedings published in this series, but also in issues of specialised scientific journals, will remain as the most important legacy of the congress.

L'UISPP a une longue histoire, à partir de 1865, avec le Congrès International d'Anthropologie et d'Archéologie Préhistorique (C.I.A.A.P.), jusqu'en 1931, date de la Fondation à Berne de l'UISPP. En 1955, l'UISPP est devenu membre du Conseil International de philosophie et de Sciences humaines, associée à l'UNESCO. L'UISPP repose sur plus de trente commissions scientifiques qui représentent un réseau représentatif des spécialistes mondiaux de la préhistoire et de la protohistoire, couvrant toutes les spécialités de l'archéologie : historiographie, théorie et méthodes de l'archéologie ; Culture matérielle par période (Paléolithique, néolithique, âge du bronze, âge du fer) et par continents (Europe, Asie, Afrique, Pacifique, Amérique), paléoenvironnement et paléoclimatologie ; Archéologie dans des environnements spécifiques (montagne, désert, steppes, zone tropicale), archéométrie ; Art et culture ; Technologie et économie ; anthropologie biologique ; archéologie funéraire ; archéologie et sociétés.

Le XVIII° Congrès mondial de l'UISPP en 2018, accueilli à Paris en France par l'université Paris 1 Panthéon-Sorbonne et avec le soutien de toutes les institutions françaises liées à l'archéologie, comportait 122 sessions, plus de 1800 communications de scientifiques venus de près de 60 pays et de tous les continents.

Les actes du congrès, édités par l'UISPP comme dans des numéros spéciaux de revues scientifiques spécialisées, constitueront un des résultats les plus importants du Congrès.

Marta Azarello

Secretary-General / Secrétaire général UISPP

# Contents

# Introduction to the session

Annick Daneels
Instituto de Investigaciones Antropológicas, Universidad Nacional Autónoma de México
annickdaneels@iaa.iia.unam.mx

The symposium 'Earthen construction technology', organised by the editors of this volume with support of CONACyT project CB254328, Mexico. It was presented as part of the theme: 'Fabrication procedures' during the XVIII Congress of the International Union of Pre- and Protohistoric Science, which took place from June 4-9 2018, at the Sorbonne University, in Paris. The general theme of the congress, 'Adaptation and sustainability in prehistoric and protohistoric societies confronted with climate change', offered a perfect context to open a discussion on construction techniques that use earth, a raw material particularly susceptible to climatic variations. The cases included do not specifically broach climate change, but always evaluate the effect of climate on the architecture and the diverse strategies developed by the ancient builders to avoid collapse and weathering, from careful selection of earthen mixes to mechanical reinforcement of walls, from stone or stucco facings to the use of organic additives to stabilize earthen renders, from surface waterproofing to enhanced drainage techniques, etc.

Though recently reevaluated because of its universal, sustainable, economical and bioclimatic values by ICOMOS (with its International Scientific Committee on Earthen Architectural Heritage - ISCEAH) and UNESCO (with the World Heritage Earthen Architecture Programme - WHEAP 2007-2017), archaeological earthen construction has traditionally been studied from the architectural point of view (evolutions in form, style and layout), on the one hand, and, on the other hand, from the focus on the conservation of existing heritage sites (where extensive excavations are limited because of the site's protected status), and on the understanding of extant vernacular building strategies. Unfortunately, vernacular building is generally geared towards domestic architecture for and by family units. The knowledge expressed in vernacular architecture does therefore not reflect the architectural and engineering know-how required to carry out the monumental achievements of civilizations like Mesopotamia, Andes or Mesoamerica, that were commandeered and backed by elite political programs. Thus, a deepened technological understanding of the finer constructive techniques of ancient monumental architecture, and its differences with domestic construction, developed worldwide over more than a 100 centuries, can only be obtained from archaeological research: extensive excavations with special attention to architectural and structural features, and their modification and collapse, coupled with typological, mineralogical, micromorphological, botanical, chemical (inorganic and organic) and mechanical studies of building materials. Most analytical techniques derive from geoarchaeology, also from engineering, but had to be adapted to analyze man-made earth mixes and their rubble (for which the Americans use the evocative term of 'adobe melt', although strictly speaking, sediment does not melt, but disaggregates).

While fabrication procedures are a common and longstanding topic of research in lithic, ceramic and metallic materials (and in a lesser degree in textile and woodcutting), architecture in Pre- and Protohistoric societies has mostly been approached in a descriptive way, with relatively little interest for its technological aspects (more so with earthen architecture as stone). Thus a technological approach is not only a novel way in the archaeology of earthen buildings, with the social and economic correlates of understanding technological procedures, but it additionally reveals a corpus of engineering and architectural know-how relevant for both the preservation of earthen heritage sites as well as the promotion of earthen architecture as a viable and economical alternative for modern construction.

Multidisciplinary approaches to earthen construction technology are a relatively new topic, acquiring a presence in archaeological literature around the 1990s, but is rapidly increasing. There is a definite snowball effect going on, as more and more results are obtained from postgraduate research and specialized publications. Mediterranean and Near Eastern archaeology is on the foreground, with mostly European teams leading the research. The French are very well represented in this volume, with references to Olivier Aurenche and Martin Sauvage for the Near East and Anne de Chazelles for the western Mediterranean, their interest in technology and operational sequences (*chaîne opératoire*), as culturally defined sets of choices that allow to understand human behavior, deriving from a longstanding research tradition arising with François Bordes and formalized by André Leroi-Gourhan and Pierre Lemmonnier; there are also research teams in Spain, England, Germany, Austria. America is still lagging behind in the aspect of multidisciplinary studies of earthen construction techniques and materials. A few US archaeologists active in North America have done micromorphological analysis with Paul Goldberg, a close collaborator of Richard Macphail. French, Spanish and German archaeologists are introducing construction sediment micromorphology and daub analysis in Central and South America, where Japanese teams are also leading important research in excavations, analysis and conservation of earthen heritage sites.

The call for papers was again answered by researchers working in Europe, Asia, America and Africa, like the preceding UISPP Memoir (Monumental Earthen Architecture in Early Societies. Technology and power display, Oxford: Archaeopress 2016). This time, the symposium consisted of 15 papers and 3 posters, presented by 34 researchers. The large number of co-authors reflects the existence of long-term projects and multidisciplinary approaches.

In this publication, only 8 papers and 2 posters are published. The 5 Mesoamerican papers will appear in extended version in the series 'Arquitectura mesoamericana de tierra' published by the Universidad Nacional Autónoma de Mexico: those on the origin of earthen architecture traditions in America (Annick Daneels), and on the sites of Xochitécatl-Cacaxtla (Mari Carmen Serra Puche), have come out in volume I, in 2019, those on Cholula (Nora Pérez), Tamtoc (Diana Zaragoza and Patricio Dávila), and Teotihuacan (Mareike Stahlschmidt, Susann Heinrich, Nawa Sugiyama, and David Carballo) are programmed in the next volumes. Unfortunately, authors Myriam Seco and Agustín Gamarra declined to send their paper on the construction sequence, building techniques and the preservation strategies of Thutmosis III's Million Year Temple at Luxor. So did the South American colleagues analyzing geoarchaeologically the construction sequences of monumental mounds in Brazil: Rafael Milheira on a 1st millennium CE mound in the Southern Pampa, and Kelly Brandão and colleagues on a 11-17th century mound in the Southwestern Amazon.

The present volume is ordered as the symposium was, going from general to particular, and more or less chronologically, starting from Near Eastern Neolithic to Mediterranean Bronze Age, then moving to America from north to south. The building techniques analyzed are mainly daub and mudbrick, though some cob is referred to.

Thus, the first paper of this volume is by archaeologists David Gandreau and Chamsia Sadozaï, and architect Sébastien Moriset, a team from CRATerre, the prestigious French institution part of the École Nationale Supérieure d'Architecture de Grenoble and UNESCO chair of earthen architecture, in acknowledgement of its 40-year trajectory. Their paper and illustrations present the major building techniques of wattle-and-daub, cob, mudbrick, and rammed earth, as well as the basic preservation strategies required when excavating earthen sites, using examples from their own research mostly in the Near East.

Emmanuel Baudoin, as part of his postdoctoral research, compares Neolithic cob and mudbrick construction in northern Mesopotamia and the Southern Caucasus, from a typological point of view, to evaluate the possible technological transfer from south to north by the 7th and 6th millennium BCE. Postgraduate researchers Paul Bacoup and Sandra Prévost-Dermarkar work on

5th millennium BCE Macedonian Neolithic architecture, that can generically be termed wattle-and daub, and through comparative paleobotanical analysis evaluate the interaction between wood selection, environment, and cultural choices. For the Upper Mesopotamian Late Chalcolithic to the Early Bronze age, Giovanna Liberotti uses architectural analysis, X-ray Fluorescence, and porosity and mechanical tests of mudbrick samples from nine buildings of five phases spanning the 4th to 3rd millennium BCE, at Arslantepe, Turkey, to assess the differences between a Syro-Mespotamian constructive tradition and that from Transcaucasian settlers.

Next is the study by Marta Lorenzon, using micromorphology, FT-IR and ESM of Middle of Late Bronze age mudbricks from three Minoan palaces in Crete, a topic, like Liberotti's, which furthers the research originating in her in-depth doctoral thesis. She addresses the effect of organic and mineral additives on the resistance of the bricks, and the thermal effect of conflagrations on the mineralogical composition of the sediments. The analysis of wood imprints and clay temper in burnt daub fragments allows Héctor Juan Fonseca to infer changes in wattle-and-daub constructive traditions between the Chalcolithic and the Bronze Age in Northern Spain, as subject of his doctoral dissertation. A second paper by Héctor Fonseca, with five co-authors and presented as poster, broaches the heritage and preservation problems of opening to public visit the earthen architecture site of El Castillar (dating to the Iron Age), 30 years after the site was excavated.

Turning to America, three papers are included. Long-time Southwestern archaeologists James Allison and Joseph Bryce describe the building systems of four types of constructions in an 8th century Pueblo village, including cob (in lumps and in layers, sometimes chinked with stones), wattle-and daub for walls and roofs, and mudbrick, reflecting a wide variety of traditions, that would ultimately be hidden behind uniform earth renders. This they can related to evidence of immigration of several groups, joining to form a new type of settlement, and also a new society. Robert Rosenswig, as part of his long-term project in Izapa on the Pacific coast of southern Mexico, uses LIDAR to survey settlement pattern in a humid tropical environment, obtaining evidence on recurrent urban layouts of earthen mounds that define the extent of Izapa's territory, and excavations to define the architectural growth of this capital of an early state in Late Formative Mesoamerica. Lastly, based on her doctoral research, Maria Torras compares in a poster the different construction techniques used to build the three major pyramids of Teotihuacan, Mexico, during the 3rd century, stamped earth vs earth contained in adobe or stone cells, evaluating the impact of monumental construction on the direct environment of the site, the subsistence of its population and the political clout of its leaders.

Some themes are recurrent in the papers and reflect the major problems associated with earthen architecture. The major one is the correct identification of the building technique. The lack of specialized courses in graduate schooling, excavation techniques switching to 'geological/stratigraphic mode' when working in earthen strata, when 'architectural/construction sequence mode' should be prioritized, and reference books based on contemporary techniques mostly derived from vernacular systems, do not prepare the professional to tackle the diversity and the complexity of solutions invented by the ancient builders over the millennia worldwide. Very often, building systems are not identified, or worse, incorrectly identified, as cases of cob labeled rammed earth (*tapia*) or mudbrick. Some terms are confusing, like *adobe*, used in American English literature to cover not only sun-dried mudbricks, but also any kind of element made of a mud-mix: floors, plasters, renders, roofs. Wattle-and-daub covers a wide variety of techniques to build the supporting wooden frames and the latticework that will hold up the daub; therefore, a much more detailed typology of the wooden frameworks is needed to understand technological and cultural differences. Mudbricks have their own problems, particularly relating to the use of molds. In cultural terms, molds are extremely significant, as mass-produced standard-sized building modules are the base of a 'modern', professional architecture and generally relate to the emergence of complex societies. The use of mold is inferred from the standardization of brick size, as no molds have been reported so far from archaeological excavations, not even on the Peruvian coast where conservation of organic material is extremely good and mudbricks exist by the million. Yet very little has been done to develop independent techniques to analyze the use of molds, like the striations or ridges on the brick due to unmolding. These are the features that Baudouin in this volume sees on plano-convex bricks, leading him to consider them mold-made, in opposition to the common appreciation that such a shape is indicative of hand-fashioned bricks. Another alternative has been proposed elsewhere, suggesting that standard-size rectangular bricks could be made in holes instead of wooden molds (Daneels, A. and Piña, J.S. 2019, 'Adobes prehispánicos fabricados en hoyo: un caso de la costa del Golfo de México' in del Cuero Ruiz-Funes, J.I., Méndez Pineda, V.M. and Huerta, S. (eds), Tercer Congreso Internacional Hispanoamericano de Historia de la Construcción, Volumen I. Madrid: Instituto Juan de Herrera: 267-276).

As the volume shows, typological analysis of architecture and of building materials are still the basic tools, but analytical techniques are plenty. As they are slow and expensive, some more than others, they are mostly only applied as part of long-term and well-financed projects; thus, building up databases will take time. The results achieved so far demonstrate they allow far-reaching anthropological interpretation of building traditions and cultures. Some are more recurrent (XRF, XRD, SEM, some FT-IR, mechanical properties, paleobotanical analysis of imprints and anthracology); the micromorphological analysis of construction material and rubble is on the rise, as it may well become the best way to distinguish between building techniques, by defining the features indicative of the amount of water in the mix and the degree and direction of compaction (Agnès Courty, Cécilia Cammas and Julia Wattez in France and Richard I. Macphail in England are pioneers in this respect). Other techniques, as GC-MS combined with IRMS, to identify residues of possible organic additives, are more rarely applied, because of the laboratory requirements and costs, and the difficulty to distinguish between natural and intentional residues, but such additives are part of the strategies to stabilize earthen architecture and their presence is therefore culturally significant (see for example Daneels et al. 2020, Bitumen-stabilized earthen architecture: the case of the archaeological site of La Joya, on the Mexican Gulf Coast, in JASREP 34).

As research teams working on archaeological earthen architecture increase in several countries, continued networking will be important. Several websites are already functioning, Réseau Terre (mostly archaeologists of the French speaking sphere), PROTERRA (including mostly architects and engineers but also some archaeologists of the Iberoamerican realm), while CRATerre and the Getty Conservation Institute have ongoing projects in heritage architecture preservation. Congresses specialized in earthen building and conservation, but that include archaeological research on the topic (though to a lesser extent) are the TERRA, organized by ICOMOS, UNESCO and CRATerre, Earth USA, SIACOT (Seminario Iberoamericano de Arquitectura y Construcción con Tierra), organized by PROTERRA, and the CIATTI by the Universidad de Valencia. Nevertheless, stronger interaction will be needed from archaeologists, so that a Scientific Committee specialized on Earthen Architecture could be warranted, as part of the UISPP domain of Technology and Economics. This was discussed at the symposium and remains as a proposal to be formalized in the 2021 UISPP at Meknès, Marocco, home to a distinguished earthen architecture tradition.

# Chapter 1

# Earthen architecture on archaeological sites: sustainability principles *vs* decay processes

David Gandreau
gandreaudavid@hotmail.com, corresponding author

Chamsia Sadozaï
ch.sadozai@gmail.com

Sébastien Moriset
sebastien.moriset@grenoble.archi.fr
Laboratoire CRAterre, Unité de recherche AE&CC
Ecole Nationale Supérieure d'Architecture de Grenoble

**Abstract**

This article highlights the sustainability of earthen architecture as revealed by many archeological sites. Despite millennia of slow deterioration processes, remarkably well-preserved sites have become sources of information and inspiration for contemporary designers because of the sustainability principles they reveal. The present quest for more sustainable ways of building will not find answers in sophisticated building techniques, but rather in what our ancestors have kept doing for centuries: building simply with the resources at hand, such as earth. But this requires close collaboration between archeologists and architects to bridge the 11 000 years gap that separates us from the first builders, from whom we can still learn a lot.

**Keywords:** Earthen construction, durability, degradation, conservation, site presentation

**Résumé**

Cet article met en évidence la durabilité des architectures de terre révélée par de nombreux sites archéologiques. Malgré des millénaires de processus de détérioration lente, des sites remarquablement bien préservés sont devenus des sources d'informations et d'inspiration pour les concepteurs contemporains en raison des principes de durabilité qu'ils révèlent. La recherche actuelle de moyens de construction plus durables coïncide entièrement avec ce que nos ancêtres ont toujours fait: construire simplement et avec les ressources disponibles, comme le permet la terre crue. Pour cela, il est nécessaire d'établir une étroite collaboration entre archéologues et architectes pour combler le fossé des 11 000 ans qui nous sépare des premiers constructeurs, dont nous pouvons réellement beaucoup nous inspirer.

**Mots-clés:** construction en terre, durabilité, dégradation, conservation, mise en valeur du site

## Introduction

From their emergence at least eleven millennia ago to contemporary times, the diverse constructive cultures associated with earthen materials have continuously spread throughout the world (Figure 1). As a result, archaeological sites presenting earth as a building material for walls, floors or roofs, are innumerable.

Some pre/protohistoric sites, still remarkably well preserved, provide the proof that earthen construction can be perfectly sustainable under certain conditions. But in many cases, with the departure of the inhabitants who maintained them, processes of degradation began which gradually led to their ruin and sometimes even their almost total disappearance (Gandreau 2017: 23-30).

Figure 1. Timeline of earthen architecture testimonies © Inclusit Design 2016

Through several examples of Near and Middle Eastern, Mediterrean and Egyptian sites built with earth, this article aims to highlight some long-lasting principles of sustainability of earthen architecture, while underlining the circumstances and processes of decay that threaten them in archaeological context.

## 2 Question of definition

'Clay', 'mud', 'earth', 'soil', '*pisé*', these terms are often used by archaeologists to characterize the structures they excavate. In fact, five principal modes of implementation, namely wattle and daub, cob, mudbrick, rammed earth, and earth renders, come in many variations, that are not always easy to distinguish in archaeological context (Houben and Guillaud 1989; Fontaine and Anger 2009). To avoid any semantic confusion, we define these main techniques as follow:

- Wattle and Daub. The building skeleton is made of wood and filled up with a woven lattice of wooden strips, then covered by an earthen mix on each side (Figure 2). Examples of this technique, made of reed instead of wood, are supposed to be the earliest human constructions. The site of Dikili Tash in Greece has evidences of wattle and daub from the Neolithic period - fifth millennium BCE (see Bacoup and Prévost-Markar, this volume).

- Cob. A wet earthen mix is laid directly on the floor in thick successive layers and the sides are cut to shape a wall (Figure 3). The wall is load-bearing by itself like a monolith. This technique can be compared to hand molded pottery with coils. Examples of cob dwellings are known in Armenia since the Neolithic period (sixth millennium BCE), in the Ararat plain (Aknashen-Khaturnakh).

- Adobe or mudbrick. A wet earthen mix is molded and sun dried to create small standard elements easy to handle (Figure 4). It has been the most widespread building system in the world for centuries. The best-known example is coming from Pre-Pottery Neolithic A (eighth millennium BCE) of the famous site of Jericho (Tell es-Sultan).

- Rammed earth. Earth is compacted in a very low humid state between two pieces of wood (Figure 5). Like cob walls, rammed earth walls are monolithic. Early evidences are scarce but some examples can be found in China.

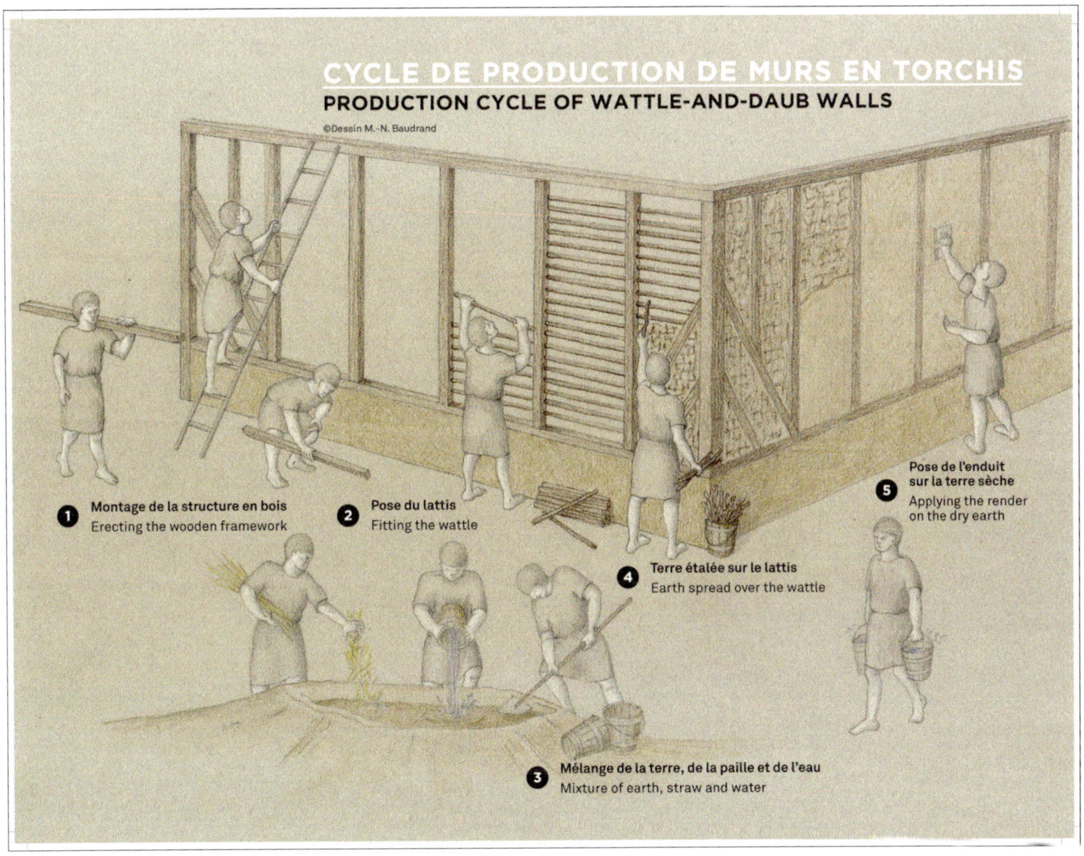


Figure 2. Wattle and daub process © M.-N. Baudrand 2016

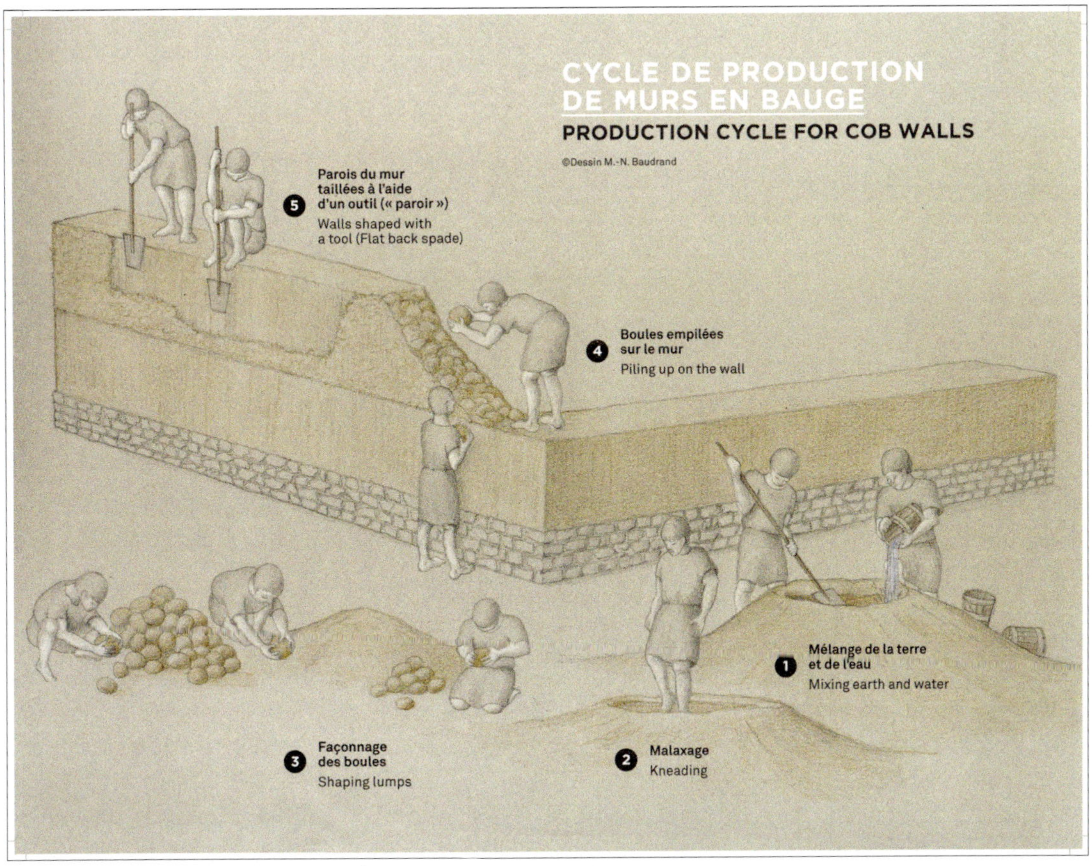


Figure 3. Cob process © M.-N. Baudrand 2016

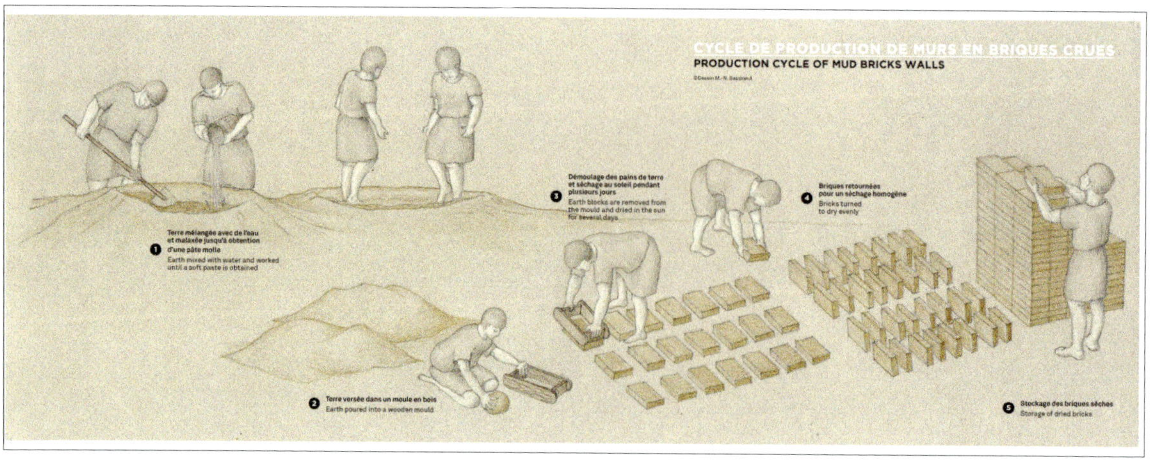


Figure 4. Mudbrick process © M.-N. Baudrand 2016

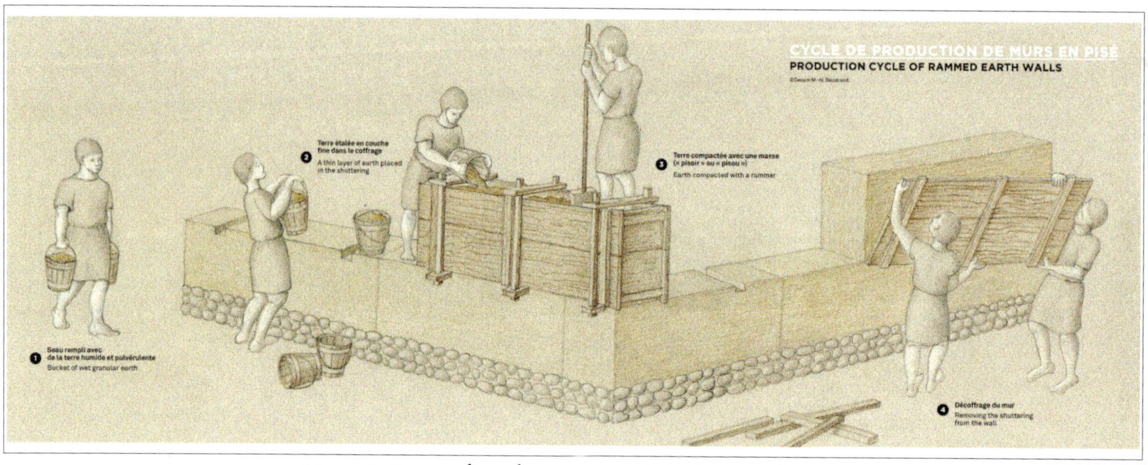


Figure 5. Rammed earth process © M.-N. Baudrand 2016

- Renders. Thin coats of earth preparation are applied on a vertical surface and can receive any type of decoration (Figure 6). Examples in Djade al-Mughara in Syria are attested since the ninth millennium BCE.

### 3 Sustainability principles

The principles of sustainability were certainly not a topic of discussion 11 000 years ago. Yet the architecture of archaeological sites teaches us real lessons of sustainability for many reasons.

The first lesson of sustainability is respect for nature. These architectures have passed through the centuries without leaving any pollution behind them, because they are made of natural materials that the construction sites provided. Earth for example was used raw; it was rarely burnt, which saved energy and ensured an impact-free return of materials to the ground.

The availability of the material, in a huge quantity with minimum transport costs, exactly under the feet of the builders, allowed the inhabitants of the past to conceive towns for several thousand people. The concept of city was born, with its large-scale vision, and was replicated for many generations thanks to the reversibility of the earthen material. The vertical stratigraphy, the fact that several occupation horizons are one on top of the other during centuries, was possible because it was easy to destroy and level old phases and re-use the same building material for the next, as it did not lose its properties. Many examples of piled-up cities are known, such as Suse in Iran which was occupied from the fifth millennium BCE until the fifteenth century CE (Amiet 1988). On a smaller scale, Ulug Depe in Turkmenistan is the only site in the region with a continuous

Figure 6. Mural painting of Djade el Mughara. © Mission archéologique de Dja'de, E. Coqueugniot

occupation from the Chalcolithic period until the end of the Iron Age (Lecomte et al. 2002), with more than 35 m of archaeological levels.

Another remarkable feature that is sometimes possible to appreciate for several millennia after the construction is the bioclimatic quality of these architectures, which avoid extreme heat or cold. The thermal inertia properties of the earth allow a slow restitution of the temperature within the walls if they are homogeneous and thick enough. Combined with a proper orientation, an adequate location on the site, and an appropriate ventilation system, these features provided comfort despite of the absence of active heating/cooling devices.

Another characteristic is the sobriety of the means mobilised. Building was done with simple tools (hoes, wooden molds) with the only energy sources available: human (and sometimes animal) for the preparation and implementation of materials and solar for drying them. This did not prevent the production of remarkable masterpieces in terms of architectural quality and decoration. Many monumental buildings are attested, such as palaces, temples, even pyramids. Indeed, a massive ziggurat was discovered in Tchoga Zanbil for the Elamite period (third millennium BCE) in Iran, comparable to the stone pyramids of Egypt. Another noticeable building is the Great Kyz Kala in Merv, in Turkmenistan, which is supposed to be a fortified house with gigantic crenulations all built in mudbricks. This building crossed the centuries and still stands up 12 m above the ground on the Bactriana plain (Figure 7).

Figure 7. The Great Kyz Kala in Ancient Merv (Turkmenistan) ©CRAterre_D. Gandreau 2015

Because of the simplicity of the materials and the means of implementation, these architectures were easy to maintain with the same resources and know-how. As a result, they ensured the resilience of their inhabitants/self-builders who carried out the entire construction process, without relying on outside resources. The earthen architecture and masonry skills those people mastered are impressive lessons for us. Even in villages such as Hili 2 in the United Arab Emirates, where there is no urban settlement comparable to those of Mesopotamia, the houses of common people are still visible today 3000 years after they erected them, from the floor to almost the ceiling 2 m above the ground. The community decided to build with thin layers of cob stabilized by thin mudbricks on top of each ring, and we still don't understand the reason for it. Another example are the Ramses granaries in Egypt, called the Ramasseum, where long and narrow rooms vaulted with mudbricks are still visible today.

Let's not forget the cultural richness of the first architectures. The simplicity of the tools and materials described above have not hindered creativity. The diversity of archaeological remains that excavations have revealed are evidence of human genius to design rich and varied cultures.

### 4 Decay processes

Archaeology has revealed the full potential of earthen architecture, but unfortunately it is also participating to its loss. Indeed, two phases of decay are visible on the lifespan of earthen archaeological remains. The first one occurs when the inhabitants abandon the building, and the second one, faster, is happening after the discovery and excavations by archaeologists.

After their abandonment, all earthen buildings, regardless of the technique employed, slowly enter in a decay phase related to erosion processes. Usually, people take away reusable and precious materials such as wooden beams or stones. Absence of roof and hard material in critical points (gutter, lintel, threshold), combined to the lack of maintenance, makes the building vulnerable to wind and water deterioration (stagnation, rain, dampness), sometimes animals squat (nest, burrow, detachment on the top). Slowly but surely, soft parts are losing cohesion and sharp features are worn away creating a big mound in the landscape. After this slow-motion process, the decaying phase is over, the mound is stabilized and the buried elements easily survive for centuries, including precious decorated surfaces (Figure 8).

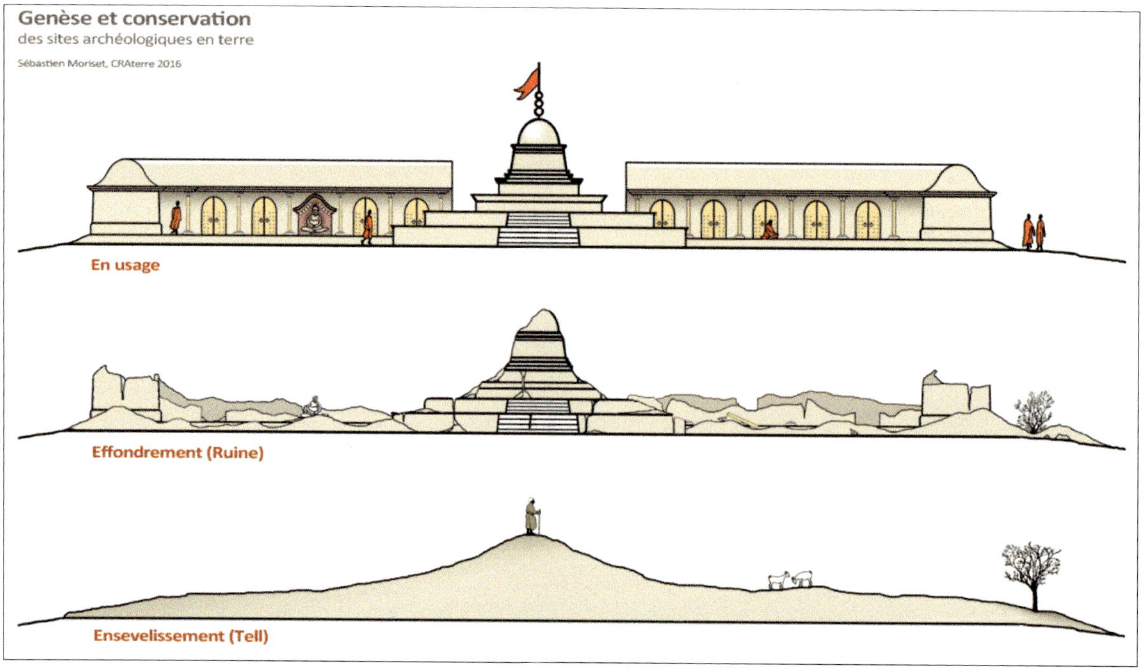


Figure 8. Decay processes after abandonment ©CRAterre_S. Moriset 2016

Figure 9. Open trench in Mary (Syria) left without any conservation measure ©CRAterre_M. Bendakir

Figure 10. Undercut pathology in Ancient Merv (Turkmenistan) ©CRAterre_C. Sadozaï 2015

When archaeologists discover a mound created by anthropic activity (also called a tepe or a tell in Middle East), a new cycle of accelerated decay is starting. Trenches are opened, walls are left exposed without roofs and trenches are dug with no way out for drainage (Figure 9).

When it is raining, water is pooling and walls are acting like sponges, absorbing all the dampness by capillarity. As a consequence, wall bases are damaged creating a weakness. This is a very common pathology, characterized by a wall profile looking like an apple core also called 'undercut' or 'coving' (Figure 10).

Other structural pathologies are observed after a long exposure without preventive conservation measures. Vertical cracks and leaning walls can, as well as the undercut, threaten the whole building's stability and can lead to a collapse. Superficial clues of degradation can also warn about the integrity of the building, such as minor cracks, material detachment or loss, and gullies (Sadozaï 2013).

Until recent times, inefficient conservation measures taken after the excavations revealed also, in some cases, major disturbances. The use of inappropriate protective materials or shelters without regular monitoring sometimes creates more damages than doing nothing (Daneels and Guerrero Baca 2015).

## 5 Learning from earthen archaeological sites

### *5.1 Conservation*

Understanding the decay processes is a key to better preserve earthen structures and possibly preserve them to future generations. Close observation of pathologies on the building and its surroundings is required to achieve a precise diagnosis (Gandreau and Sadozaï 2015). The conservation principles applied to earthen archaeological remains are well-known and are based on common sense, for example do not allow water to stagnate near the structures and impregnate the walls by capillarity, or do not leave fragile elevations exposed to wind, rain and frost without protection.

This sounds simple, but preserving an excavated site is a great challenge and can hardly be achieved once the remains are exposed. Conservation strategies need to be integrated in the archaeological campaigns, with time, financial resources and knowledgeable experts entirely dedicated to the task. Conservators should work alongside with archaeologists to apply the appropriate emergency measures required to stabilize the whole excavated sector until all the stakeholders agree upon a holistic strategy (Sadozaï and Gandreau 2016).

In theory, for any new archaeological project a conservation strategy must be thought out before starting. Many governmental authorities have understood this today and ask for realistic conservation plans to be presented before issuing excavation permits. Several archaeologists have been denied the right to dig because they could not meet the required conservation guarantees. The conservation program can of course vary depending on the findings, but it helps at least to determine in a specific time frame the appropriate conservation techniques to use, as well as the roles and responsibilities of each partner.

### *5.2 Contemporary architecture*

Does it make sense to oppose these thousand years old ruined architectures versus contemporary ones? It is a trend for architects today to reconnect their contemporary production to the vernacular architecture of the region where they build. This seems logical today in a context where it has become essential to save resources by reducing the transformation and distribution circuits of building materials by using locally sourced materials and relying on the human resources of the place. This is exactly what our ancestors tried to achieve for several millennia. What we see as outdated productions of the past often appear to be extraordinary sources of inspiration for today's architects. The idea is not to duplicate ancient models, but to draw inspiration from various parameters. Which type of soil was used for which building parts? How was the soil processed and turned into buildings parts? Which tools and social organization allowed achieving building so simply and so fast? What features guaranteed thermal comfort in the absence of electricity? Many more questions can be asked. A European Union funded research programme called VERSUS (www.esg.pt/versus) listed in 2014 all the lessons that can be drawn from vernacular architecture to explore new ways to design contemporary sustainable buildings (Guillaud et al. 2014). It provides a methodology for scholars, students and professionals to easily draw the lessons from past architectures, regardless of their age, that can be reused in a contemporary way (Sanchez et al. 2018).

On an archaeological site, the link between ancient and contemporary architecture can easily be done through a site presentation strategy that requires erecting new buildings. The necessity of exhibiting artefacts in situ in a visitors' centre for example offers the possibility of building a contemporary structure with modern features inspired by ancient techniques. The project of Mari in Syria is a good example of a site museum built in earth with modern features (Figure 11).

Figure 11. Mari Visitors' center (Syria) ©CRAterre_M. Bendakir 2010

## 6 Conclusion

The archaeological remains of earthen constructions are rarely considered by contemporary builders as sources of information that could guide their practice. The fragility that the ruins show us immediately destroys the idea that these architectures bear lessons that have remained relevant over the millennia. However, conservation has brought many archaeologists closer to architects who are committed to contemporary issues of sustainable development, for whom the use of eco-responsible materials such as earth is obvious. The dialogue between archaeologists and architects has connected very old buildings to contemporary productions, around similar materials and techniques, simply facilitated by the tools available today. The extraordinary state of conservation of certain earthen architectures with elevations of several meters, after millennia of slow decay, proves the durability of this material, and its ability to respond to the principles of sustainability. Most of the solutions are in the past, it is for us to dig them up, revitalize them and adapt them to our contemporary needs, creating a connection between those who understand the past (archaeologists) to those who design the future (architects).

## Bibliography


Amiet, P. (1988) *Suse : 6000 ans d'histoire*. Paris: Editions de la Réunion des musées nationaux.

Daneels, A. and Guerrero Baca, L.F. (2015) 'Archaeological earthen sites in America. Study and preservation' in Gandreau, D. and Sadozaï, C. (eds) *Actes du séminaire international : la conservation des architectures de terre sur les sites archéologiques. Nouvelles pratiques et perspectives.* CRAterre-ENSAG, 26-28 mai 2014. Villefontaine : CRAterre, International scientific workshop 01: 92-103.

Fontaine, L. and Anger, R. (2009) *Bâtir en terre. Du grain de sable à l'architecture.* Paris: Belin, Cité des sciences et de l'industrie.

Gandreau, D. (2017) Patrimoine archéologique en terre et développement local. Enjeux interdisciplinaires et perspectives de formation. PhD Thesis. Grenoble: Université Grenoble Alpes, ENSAG.

Gandreau, D. and Sadozaï, C. (dirs.) (2015) *Conservation des architectures de terre sur les sites archéologiques: nouvelles pratiques et perspectives.* Actes du séminaire international, ENSAG, 26-28 mai 2014 [online].Available at: https://craterre.hypotheses.org/795 [accessed August 1 2020].

Guillaud, H., Moriset, S., Sanchez Munoz, N., Sevillano Gutierrez, E. and Correia, M. et al. (2014) Versus: lessons from vernacular heritage to sustainable architecture [online]. CRAterre-ENSAG. Available at: https://hal.archives-ouvertes.fr/hal-01159770 [accessed: June 13 2019].

Houben, H. and Guillaud, H. (1989) *Traité de construction en terre.* 3rd edn. Marseille: Parenthèses.

Lecomte, O., Francfort, H.P., Boucharlat, R. and Mamedow, M. (2002) 'Recherches archéologiques récentes à Ulug dépé (Turkménistan)', *Paléorient*, 28(2): 123-132.

Sadozaï, C. (2013) 'Préservation de l'architecture de terre en Asie centrale : l'exemple du site protohistorique d'Ulug Dépé (Turkménistan)' in Bendezu-Sarmiento, J. (dir.) *L'Archéologie française en Asie centrale. Nouvelles recherches et enjeux socioculturels.* Cahiers d'Asie Centrale n 21/22. Paris: Édition-Diffusion De Boccard: 623-630.

Sadozaï, C. and Gandreau, D. (2016) 'Conserver après la fouille : la question des structures en terre crue à partir d'exemples en Asie centrale' in Stucky, R.A., Kaelin, O. and Mathys, H.P. (eds) *Proceedings of the 9th International Congress on the Archaeology of the Ancient Near East:* June 9-13 2014, University of Basel. Basel: Harrassowitz, vol. 3: 781-789.

Sanchez Munoz, N., Moriset, S., Mecca, S., Dipasquale, L., Achenza, M., Mileto, C., Vegas Lopez-Manzanares, F., Correia, M. and Carlos, G. (2018) 'Transmission de la méthode VerSus 'from Vernacular to Sustainable' aux étudiants et enseignants d'architecture: retour d'expérience' in *Terra Education III.* Grenoble: CRAterre, AE&CC.

## Chapter 2

# Spread and Independent Technical Invention of the Earthen Material in the Southern Caucasus and Northern Mesopotamia during the sixth Millennium BCE

Emmanuel Baudouin
Attaché Temporaire d'Enseignement et de Recherche (ATER), Université Toulouse Jean Jaurès,
UMR 5608 TRACES, Maison de la recherche
emmanuel.baudouin1@gmail.com

**Abstract**

The existence of relationships between Southern Caucasus and Northern Mesopotamian communities was inferred since the 1960s. Since 2008, international archaeological teams have focused on the Southern Caucasus in order to verify these hypotheses and understand the 'neolithisation process' on the periphery of the Near East. Here I will focus on the evolution of cob and mudbrick. An evolution of these earthen material techniques during the sixth millennium BCE is proposed by means of a re-examination of stratigraphic contexts, bibliographical data and also by recent discoveries in these regions. The spread of cob could be traced across Northern Mesopotamia since the end of the seventh millennium BCE, and then towards Eastern Anatolia and Southern Caucasus during the sixth millennium BCE. Simultaneously, the molded plano-convex mudbrick appeared in the Shulaveri-Shomu culture in the sixth millennium BCE. This special technique seems to be a local and independent innovation, according to other architectural characteristics.

**Keywords:** Neolithic, Earthen architecture, Sun-dried mudbrick, Cob

**Résumé**

Les relations culturelles entre les communautés du Caucase et celles de Mésopotamie du Nord sont attestées depuis les années 1960. La reprise récente des investigations dans le Caucase a pour ambition de comprendre le processus de néolithisation en marge du développement des communautés mésopotamiennes. Je concentrerai mon propos sur l'évolution de la bauge et de la brique crue. Une évolution des ces techniques de construction au sixième millénaire AEC peut être proposée par une révision de la stratigraphie, de la bibliographie et par l'apport de données inédites. La question de la diffusion de la bauge au septième millénaire, du Nord de la Mésopotamie vers l'Anatolie orientale et le sud du Caucase durant le millénaire suivant, peut être posée, alors qu'en même temps, dans la culture de Shulaveri-Shomu, apparaît la brique crue moulée plano-convexe. Cette technique particulière semble y être une innovation locale et indépendante, hypothèse confortée par d'autres éléments architecturaux.

**Mots-clés:** Néolithique, architecture en terre, brique crue, bauge

## 1 Introduction

At the end of the seventh millennium BCE, the Syro-Mesopotamian were in an advanced stage of the Neolithic 'way of life' and three main cultural horizons can be identified in this period (Figure 1). The first is the Hassuna culture that appeared at the middle of the seventh millennium BCE in Iraqi Jezirah. Proto-Hassuna sites have been identified in the Upper Khabur (Le Mière 2000), but no proper 'Hassuna' sites have been discovered, while Hassuna-Samarra sites are attested in Upper Tigris at Hakemi Use (Tekin 2011), and lasted until the end of the seventh millennium BCE.

The second is the Samarra culture (6300-5800 cal. BCE), which is known in Eastern Jezirah and Central Mesopotamia. Based on ceramic data, research shows that Samarra represents a geographical extension of Hassuna in the Jezirah (Forest 1996: 36; Hole 1977: 12). This expansion could be explained by a movement of the Hassuna population to the south, in Central Mesopotamia, result of a demographic pressure applied by the arrival of the Halaf communities (Sauvage 2001: 426).

This opinion must be qualified (Butterlin 2018: 107) because of the unreliability of the stratigraphic contexts, especially for the Zagros foothills as at Choga Sefid (Oates 1987: 166).

The third is the Halaf culture that spread over a large area stretching from Cilicia to the east, as far as the Zagros foothills and occupying essentially all the High Jezirah. Some archaeologists suggested that Halafian communities originated from the mountains located between the Van Lake and the north Iraqi steppe (Mallowan 1936; Mallowan et al. 1935; Mellaart 1970). However,

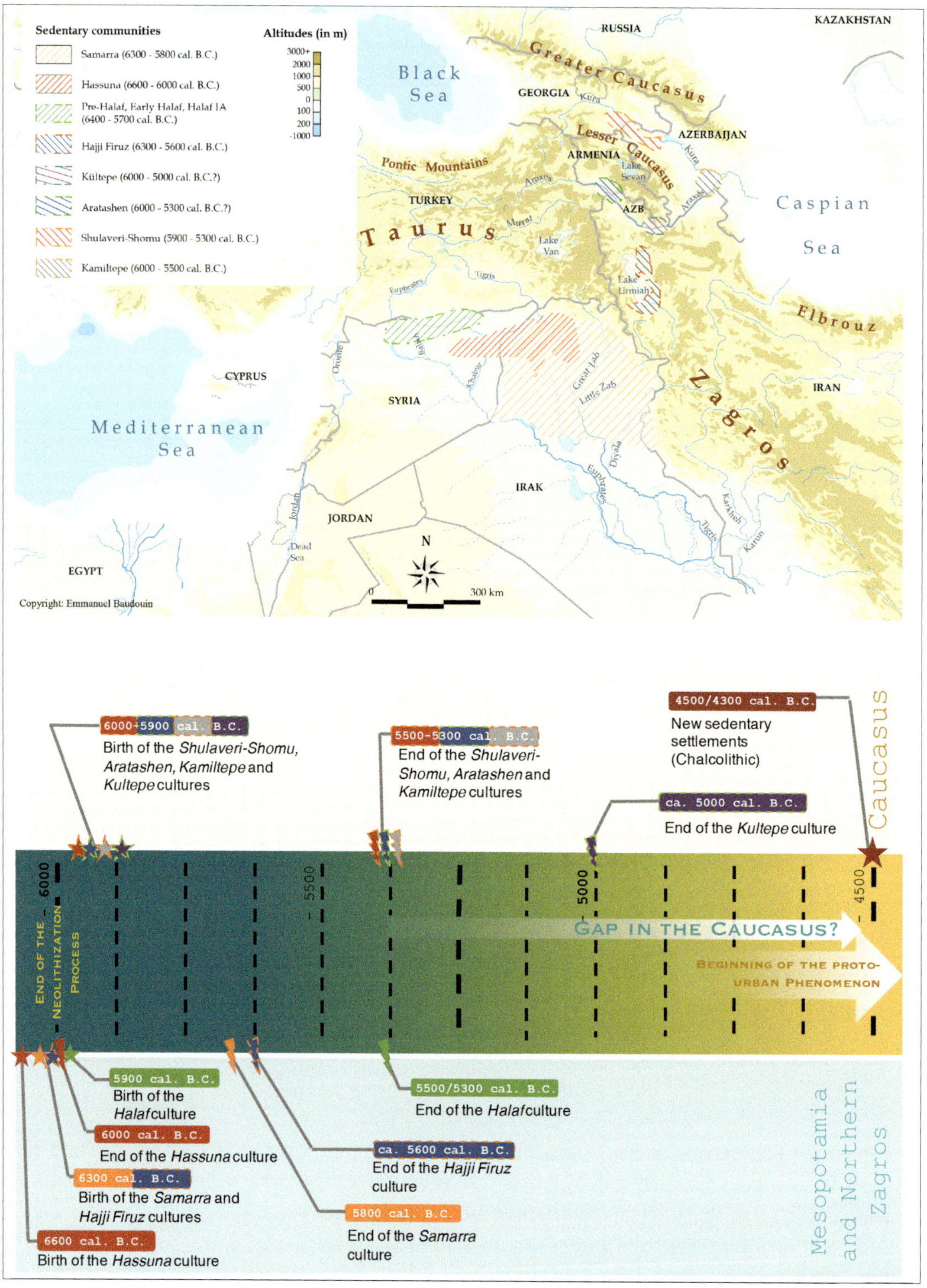


Figure 1. Map and chronology of the Northern Mesopotamia and Southern Caucasus communities from the end of the seventh millennium to the middle of the fifth millennium BCE.

other archaeologists trace the birth of Halaf around 5600 cal. BCE in High Jezirah, in the Balikh and Khabur region. Two main assumptions contrast about Halaf's origin and development: 1) the appearance of Halaf populations from Van Lake and Eastern Jezirah areas; 2) a local origin replacing previous cultures (Proto-Halaf, Proto-Hassuna, Hassuna, Samarra). Recent research (Campbell 2007) enabled to place the beginning of Halaf at the end of the seventh millennium BCE (Molist et al. 2007: 4; Akkermans et al. 2014: 30). But the origin of this culture still remains controversial (Akkermans 2000 2010; Breniquet 1996: 63–64) just as its disappearance at the end of the sixth millennium BCE.

Outside of Mesopotamia, the Hajji Firuz culture, that developed in Northern Zagros, close to the Urmiah Lake, is still poorly known and it is mainly documented by the eponymous site dated around 6300-5600 cal. BCE (Lawn 1974: 222; Stuckenrath 1963: 90).

On the other hand, the first studies on the sedentary communities of the sixth millennium BCE in the Southern Caucasus began in the middle of the XXth century. Excavations were carried out almost simultaneously in the Middle Kura Valley, Azerbaijan, and in the Kvemo-Kartli Plain, Georgia, where archaeologists identified a unique culture, so-called Shomu-Shulaveri or Shulaveri-Shomu (Dzhavakhishvili et al. 1975; Kiguradze 1986; Kushnareva 1997; Narimanov 1965 1987). At the same time, excavations were undertaken in the Ararat Plain, on the Araxes Valley, which revealed the archaeological remains of a contemporary occupation to the south of the Lesser Caucasus, a regional variant of Shulaveri-Shomu (Kiguradze 1976; Badalyan et al. 2014). Previously, other research had been undertaken at Kültepe, Nakhichevan (Abibullaev 1959 1982), where the discovery of Halaf ceramics has enabled archaeologists to hypothesise relationships with contemporary cultures in Mesopotamia (Abibullaev 1959; Narimanov 1987).

From the second half of the XXth century, the issues focused on the origin of Shulaveri-Shomu and especially on the neolithisation process in the region and three main hypotheses were made:

a) A 'colonization' by communities from Mesopotamia (Munchaev 1975);

b) Relationships and exchanges between Syro-Mesopotamian communities (Hassuna, Samarra, Halaf) and those of Southern Caucasus (Kiguradze 1986; Kushnareva 1997);

c) An independent and local development (Lisitsyna et al. 1977; Niebieridze 1986).

The resumption of research since 2008 in the Middle Kura Valley, at Mentesh Tepe (Lyonnet et al. 2012 2016 2017; Lyonnet and Guliyev 2012), Göy Tepe (Guliyev et al. 2009 2012 2014; Nishiaki et al. 2015a), Haci Elamxanlı Tepe (Nishiaki et al. 2015b) and Kiçik Tepe, in the Kvemo-Kartli Plain, at Aruchlo (Hansen et al. 2006 2007; Hansen et al. 2012) and Gadachrili Gora (Hamon et al. 2016) and in the Ararat Plain, at Aratashen (Arimura et al. 2010; Badalyan et al. 2007), Akhnashen-Katunarkh (Badalyan et al. 2014; Badalyan et al. 2007) and Masis Blur (Martyrosyan-Olshansky et al. 2013), has set as an objective questioning these assumptions.

Simultaneously, the identification of contemporary sites in the Mil Plain has enabled to identify another cultural horizon, known as the Kamiltepe culture (Lyonnet et al. 2012). In spite of the regional variants, it is customary to define a unique Neolithic culture, that of Shulaveri-Shomu. However, even if these communities homogeneously adopted a sedentary way of life based on agriculture and animal herding (Benecke 2017) from the beginning of the sixth millennium BCE, one of the main regionalised traits of these communities is represented by architecture. Technological differences conduct us to distinguish four 'cultural' entities: Shulaveri-Shomu, in the Kvemo-Kartli Plain, the Middle Kura Valley and the Karabagh Plain; Aratashen, in the Ararat Plain; Kültepe, in the Nakhichevan; and Kamiltepe, in the Mil Plain.

The purpose of this article is to define exchanges of know-how and techniques between communities of the Syro-Mesopotamian Basin and Southern Caucasus throughout the sixth millennium BCE. I will focus on the development of sun-dried mudbrick and cob. We posit two main hypotheses: technical transfer or autonomous inventions (Leroi-Gourhan 1945). We will see whether it is conceivable and relevant to define a 'center' and a 'periphery' or rather propose more diffuse relationships between the communities of these regions (Breniquet 2016: 208; Lombard et al. 2004). This study is based on the examination of 56 sites and more than 200 instances of mudbrick or cob inventoried and individually numbered in table 1 (inserted at the end of the chapter), intending to keep these records as objective as possible due to the ambiguity and variety of terms used in earthen architecture (Aurenche et al. 2011).

## 2 Methodology, *chaîne opératoire*, identification

The study of building materials provides a better understanding of the mechanisms associated with technological changes and emphasises the choices made by certain populations, thus making it possible to identify 'cultural' entities (Sauvage 2001: 428). Over the last three decades, specialized studies have clarified the archaeological study of mudbrick, *pisé* or rammed earth, and cob (Figure 2 ; Aurenche et al. 2011; Chazelles et al. 2003 2011; Guillaud et al. 2007; Roux et al. 2010); each represents a different type of earthen construction, featuring different *chaînes opératoires.*

### *2.1 'Terre massive': pisé vs cob*

The notion '*terre massive*' (stacked mud) has been introduced by C.A. de Chazelles indicating a building technique which consists of setting the earth directly where the wall is planned to stand. The two main techniques are cob and *pisé*. In *pisé*, earth is compacted in a wooden formwork with a rammer (Doat et al. 1979: 13–92). This technique is not attested historically before the Roman period (Aurenche et al. 2011: 22; Chazelles 1997: 95).

Like *pisé*, cob is shaped directly on the location of the wall. As it's made of mud, a drying time is necessary between each layer, in order to avoid any defect. J.C. Roux and C. Cammas (2010: 222–23) identify five implementation techniques. Two of them seem to be used frequently in Near Eastern and Southern Caucasus: cob in lumps and in layers. In the Near Eastern, the earliest instances are from the ninth millennium BCE and appear simultaneously in the Levant, at Gilgal (Noy 1985; Noy et al. 1980), in Eastern Jezirah, at Nemrik 9 (Kozlowski et al. 1990: 352), in the Zagros foothills, at Tepe Guran (Melgaard et al. 1963: 110) and in the Middle Euphrates and Mureybet (Aurenche 1977).

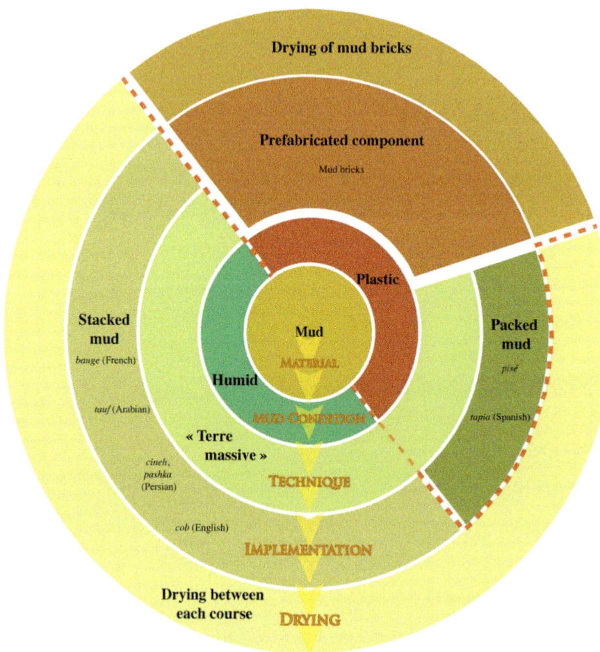


Figure 2. Summary diagram of implementation using earthen material (after Baudouin et al. 2018: 60, Figure 11).

### *2.2 Prefabricated components (mudbricks)*

Unlike stacked mud, prefabricated components are made in series and in advance. After drying, they are bonded with mortar in superimposed courses to make the wall; this technique is much faster than cob. Mudbricks can be shaped by hands or made in a wooden mold. This second technique allows to standardize both shape and size of the mudbricks and to increase the production rate (Aurenche 1981: 66). It should be noted that duly documented examples of molded mudbricks are rare and it is often only in terms of the regularity of the modules and the length/width ratio that it can be deduced that they are mold-made (Baudouin, tbp 2020).

For the molded mudbricks, a thorough dismantling of the wall during the excavation enables to observe two features of the *chaîne opératoire*: 1) the removal of the excess of mud by scraping leaves circular or longitudinal marks on the surface (Baudouin 2017: 11–12, 159–160); 2) the removal of the mold leaves often a tiny ridge clearly visible at the top edges of the mudbricks (Aurenche 1981: 18, 64; Baudouin 2017: 159-160, fig. 13).

At the current state of knowledge, the first mudbricks shaped by hand appear in different places in the Near East at the beginning of the ninth millennium BCE (Aurenche 1993: 84; Sauvage 1998: 192). Their shapes are diversified: plano-convex at Nemrik 9 (Kozlowski 1989: 27; Kozlowski et al. 1990: 11, pl. 1, 355) or also cigar-shaped at Jericho (Kenyon 1957: pl. 11B, 33A) and Choga Bonut (Kantor 1978: 191, pl. IIa). On the other hand, the first molded mudbricks appeared in the second half of the eighth millennium BCE in the Middle Euphrates Valley at Cafer Hüyük (Aurenche et al. 1985: 13–15), at Gritille (Aurenche et al. 1988: 5), at El-Kowm 2 (Stordeur 2000: 37–38), at Qdeir (Contenson 1985: 338, 340) and at Tell Halula (Molist et al. 2014: 108).

## 3. Results

### *3.1. Mesopotamian and Northern Zagros communities*

#### *3.1.1 Cob*

In Jezirah, at Telul eth-Thalathat n°197 (Proto-Hassuna, level XV), Japanese scholars identified elongated and irregular lumps, piled up without mortar (Fukai et al. 1970: 7, 22). My recent study[1] of the photographic archives of this level shows the regularity of courses (around 8-9cm) for some walls and the irregularity of lumps on each course.

At Tell Hassuna n°87 (Hassuna, levels I-II), 'lumps of mud of various size' (Lloyd et al. 1945: 273) are arranged in a wet state on the location of the wall. After drying, both sides of the wall were smoothed to homogenize the surface. At Tell Sabi Abyad n°124,126, lumps seem used from level 8 (Op. I, Pre-Halaf 6125-6075 cal. BCE), excavators recognized 'layers of slabs' (Akkermans et al. 2014: 36), as at Tell Arpachiyah n°7 (Halaf 6400-5400 BCE) where it is used for the construction of tholoi (Mallowan et al. 1935: 25). Layers of cob are attested at Tell Sabi Abyad n°132-133 (Op. I, Early Halaf, level 6) and namely a wall made of different layers of mud with different colours and textures; this technique continues at least up to level 3B, where excavators describe a similar alternation of layers (Verhoeven et al. 1996: 44, 94). At Yarim Tepe II (Halaf IA, level VI), the wall of tholos 41 n°201 consists of 'clay layers', joined with a yellowish mortar (Merpert et al. 1976: 45), close to the technique used for the circular buildings 44A and 44B n°202 (Merpert et al. 1977: 89). In the Northern Zagros, at Hajji Firuz n°76-80 (Hajji Firuz, levels J to A2), earthen beds are arranged in successive layers along the entire length of the wall (Voigt 1983: 33). This technique seems to persist in the region during the Dalma period where layers of cob have been identified at Dalma Tepe n°52 in levels 5-4 (Hamlin 1975: 113).

Nevertheless, many architectural features remain unspecified. At Tell Sabi Abyad n°125,128 (Op. I, Pre-Halaf, level 7B), the architectural elements appear as homogeneous 'mud blocks', maybe similar to the 'slabs' described in levels 8 and 7B (Akkermans et al. 2014: 37, 45, 52). At Mattarah n°109 (Hassuna, levels I-II), the walls are reported as packed mud (tauf), without additional precisions

[1] I thank Professor Y. Nishiaki (The University Museum, The University of Tokyo) for giving me access to these archives. This research was made thanks to the Japan Society for the Promotion of Science (JSPS) as part of a postdoctoral fellowship at the University of ToKyo.

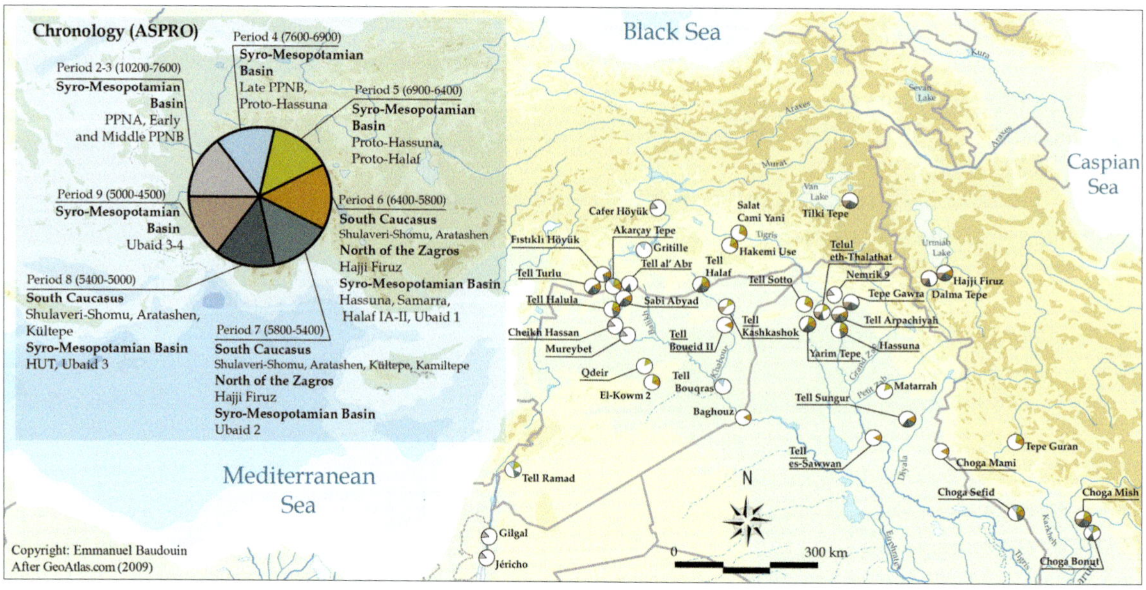


Figure 3. General map of Northern Mesopotamia with the sites mentioned in the text.

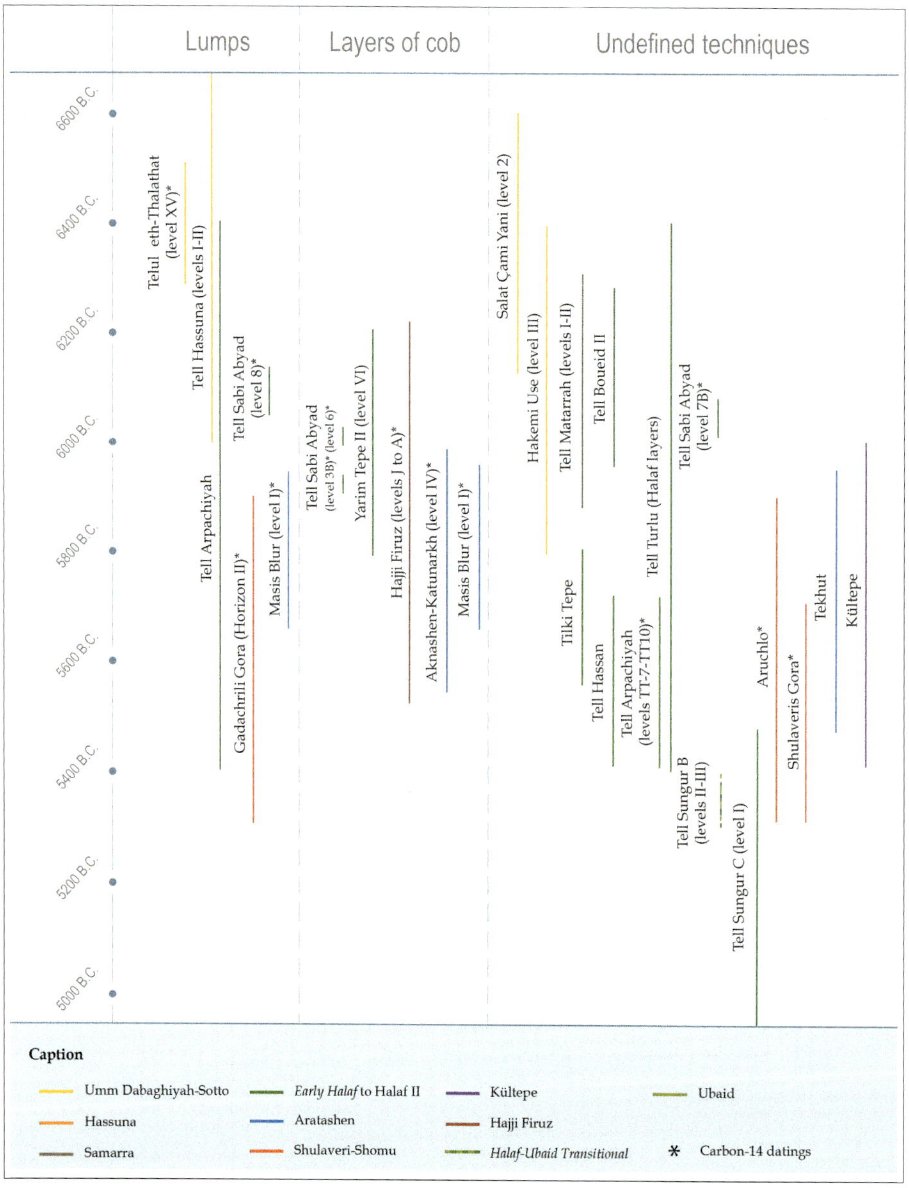


Figure 4. Summary diagram of the attestations of the stacked mud technique (seventh-sixth millennium BCE)

Figure 5. Telul eth-Thalathat II (level XVb): a-b.) Wall in cob in squares O-VIII, view from the west; c-d.) Walls in mudbricks in squares N-VIII, N-IX, view from the south-west. Japanese archaeological excavations at Telul eth-Thalathat (campaign 1976), unpublished archives of the University Museum, The University of Tokyo.

(Braidwood et al. 1952: 6). At Tell Boueid II n°39 (Proto-Hassuna/Pre-Halaf), at Tell Arpachiyah n°8 (Halaf II, TT7 to TT10), at Tilki Tepe n°198 (Halaf II), at Salat Cami Yani n°141 (level 2) and at Hakemi Use n°89 (Hassuna or Samarra, level III) buildings are made of cob, though incorrectly called *pisé* by the authors (Suleiman et al. 2002: 6; Mallowan et al. 1935: 25; Chataigner 1995: 59; Tekin 2011: 152–53; Miyake 2011: 131), or called tauf, as at Fıstıklı Höyük n°53 (Halaf I, level II) and at Tell Hassan n°92 (Halaf II) (Bernbeck et al. 2003: 164; Jasim 1985: 164). At Tell Turlu n°199 (Halaf), a similar material may be used though no precisions are available (Breniquet 1987: 113). Finally, data at Tell Sungur n°189-194 shows the use of cob in Halaf-Ubaid Transitional (HUT) and in Ubaid 2-3. At Tell Sungur B (HUT, levels I to III), the walls of buildings B1 and B2 are built with 'tauf' (Fuji 1981: 184, 190), just like at Tell Sungur C (Ubaid 2-3, level I) (Fuji 1981: 188, 190, 191; Jasim 1985: 156).

In addition, the recent recovery of the archives of the sites of Telul eth-Thalathat and of Tell Kashkashok shows that cob initially identified by the excavators in the Proto-Hassuna levels has to be questioned. Indeed, at Telul eth-Thalathat, some walls of level XVb (squares NVIII-NIX) seem built with mudbricks, the courses being apparently regular (around 8-10cm) with some modules being visible, as well as the mortar between these bricks and the interior plaster still preserved. At Tell Kashkashok (level 3, square G7), walls preserved on at least one course seem made of mudbricks (possibly 30x25cm) and not in *tauf* (Matsutani 1991: 11, 13) (Figure 5 c-d, Figure 6).

Figure 6. Tell Kashkashok (level 3): a-b.) Wall in mudbricks in square G7, view from the south-west; c-d.) The same wall, view from the south-east. Japanese archaeological excavations at Tell Kashkashok (campaign 1988), unpublished archives of the University Museum, The University of Tokyo.

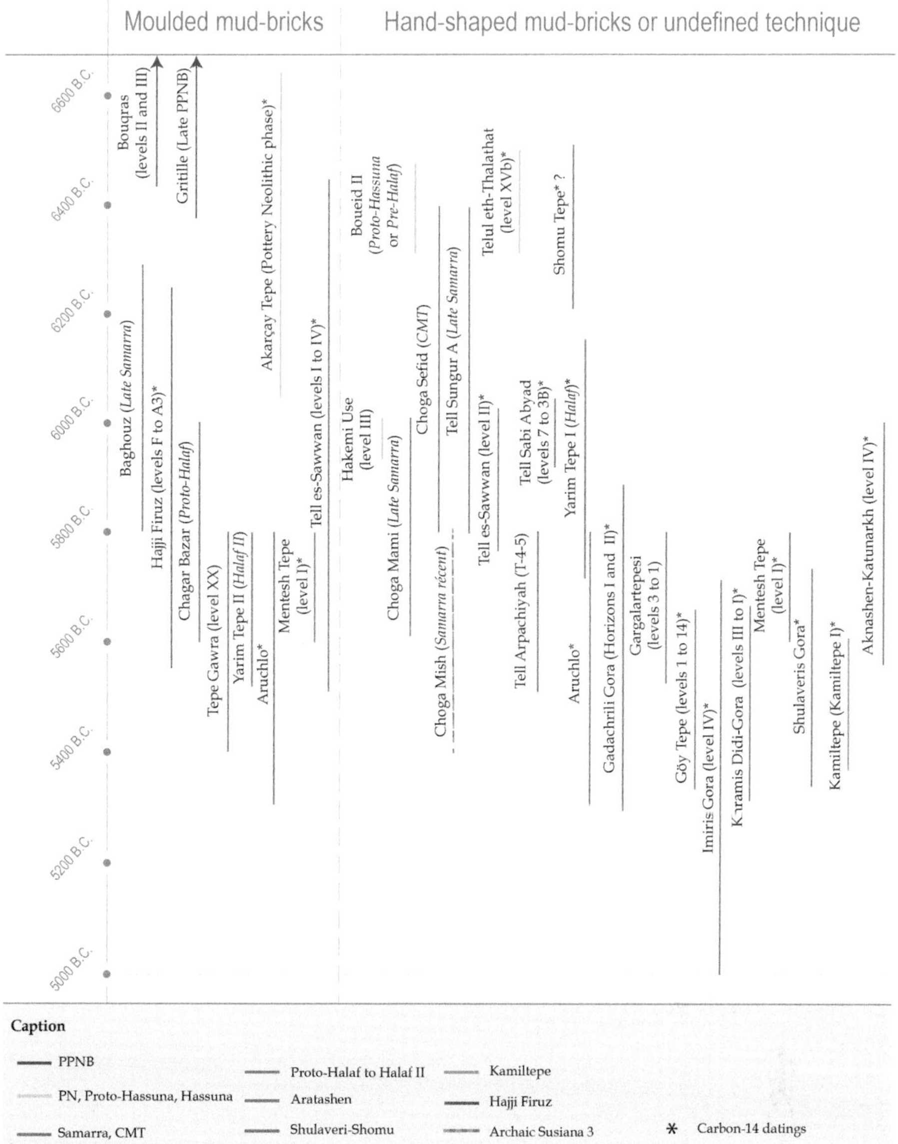


Figure 7. Summary diagram of the attestations of the mud-bricks shaped by hands and molded mud-bricks (seventh-sixth millennium BCE)

### 3.1.2 Mudbricks

Around 6000 BCE, in Eastern Anatolia, at Hakemi Use n°90-91 (Hassuna or Samarra, period III), the size of mudbricks is not standardized (Tekin 2011: 152), their length/width ratio varying between 1 and 2:1

Between 6400 and 5800 BCE, in Western Jezirah, hand-shaped mudbricks and cob are used concurrently at Tell Sabi Abyad (see above). During the Pre-Halaf phase n°129-131 (level 7), the mudbricks used in buildings 7.5 (Akkermans et al. 2014: 57) and 7.11 (Akkermans et al. 2014: 61) are close to square (ratio: 0.9-1.4:1). This type of mudbrick is used until level 3B n°137-140 where it becomes smaller as in *tholos* I (Verhoeven et al. 1996: 96) and *tholos* N/AE (Verhoeven et al. 1996: 97). Within a century (between 6020 and 5905 cal. BCE), there is a gradual decrease in the size while the length/width ratio is constant (Figure 9).

At Tell Boueid II n°40 (Proto-Hassuna or Pre-Halaf), rectangular mudbricks are used occasionally and are not standardized (Suleiman et al. 2002: 6). At Yarim Tepe I n°200 (Halaf), the large size of the elongated 'mudbricks', 90x15x6cm (Munchaev et al. 1973: 6), suggests that it is rather cob. In Eastern Jezirah, at Tell Arpachiyah n°9-12 (Halaf II, levels TT4-5), the mudbricks are irregular

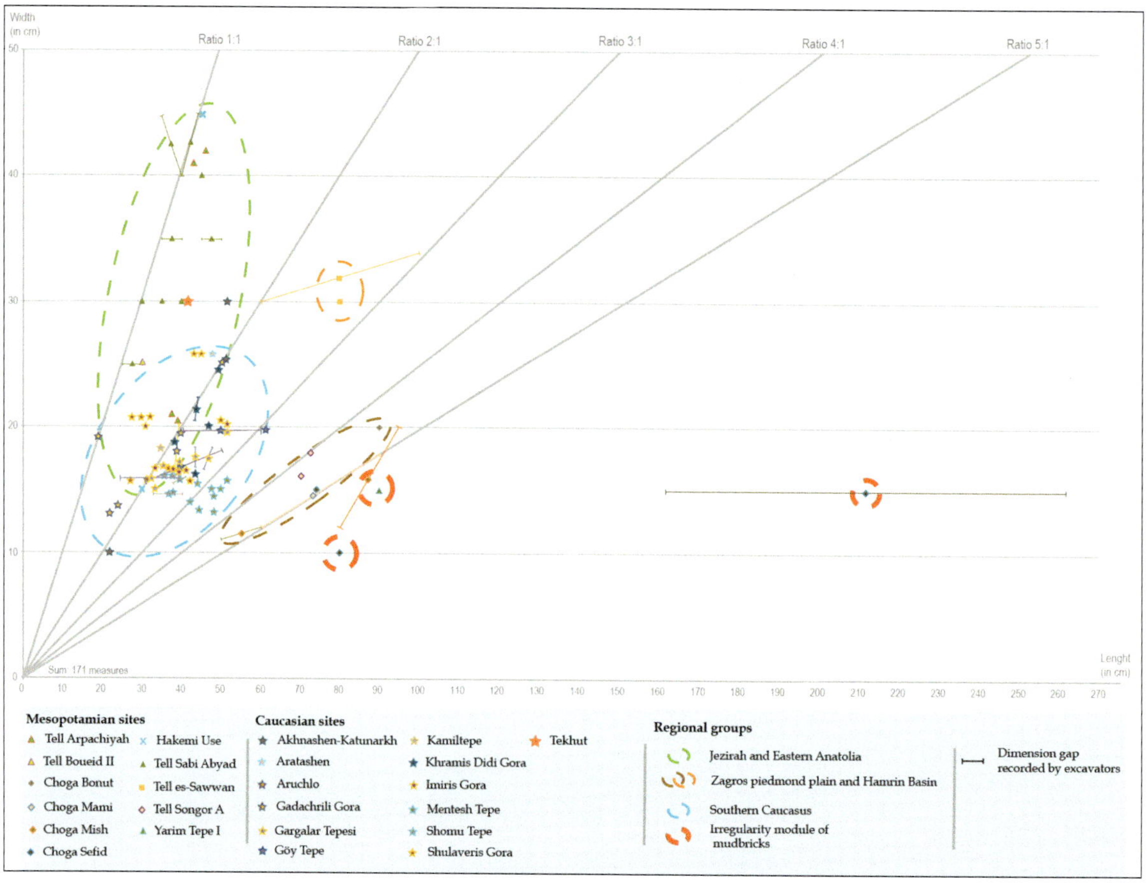


Figure 8. Summary diagram of the attestations of the mud-bricks shaped by hands in Northern Mesopotamia and Caucasus (seventh-sixth millennium BCE)

| Level | Date | Building | Brick module (in cm) |
|---|---|---|---|
| 8 | 6125-6075 BC | 8.2 | 120 x 35 x ? |
| | | 8.3 | 100 x 35 x ? |
| 7 | 6020-5995 BC | 7.5 | 45-50 x 35 x 6-8 |
| | | 7.11 | 40-44 x 40-45 x 4-7 |
| 6 | 6010-5995 BC | 6.1 | 45 x 40 x 10 |
| | | 6.13 | 40 x 40 x 4-6 |
| 5 | 6000-5945 BC | 5.1 | 40 x 30-35 x 8 |
| | | 5.2 | 30 x 30 x 12 |
| 4 | 5980-5925 BC | Tholos en Q15 | 25-30 x 25 x 8-10 |
| | | Tholos en Q15 | 40-45 x 30-35 x 8-10 |
| 3B | 5940-5905 BC | Building I | 35-40 x 35 x 10 |
| | | Tholos S | 35 x 30 x 8 |
| | | Tholos I | 30 x 25 x 8 |
| | | Tholos N/AE | 35 x 30 x 8 |

Figure 9. Summary table of the evolution of mud-bricks sizes at Tell Sabi Abyad (Syria) (Op. I) between levels 8 and 3B (6125-5905 BCE), based on Akkermans et al., 2014 (after Baudouin et al., 2018: 66, fig. 16).

(Mallowan et al. 1935: 16) and the production is divided into two categories (Aurenche 1981, tab. 6): large and almost square mudbricks, with a ratio around 1:1, representatives of Halaf (Tell Sabi Abyad), and smaller mudbricks, with a ratio around 2:1, conforming to the standardized molded mudbricks of the same period.

In Central Mesopotamia, while the use of molded mudbricks is becoming widespread in Samarra (5800-5400 cal. BCE), some mudbricks are still shaped by hand: at Tell es-Sawwan n°151-152 (PN(?), level II), the length varying between 60 and 100 cm, for a ratio between 1.8 and 3.3:1 (Wahida 1967: 172). The weight of these bricks is estimated around 47 kg for the largest ones. Such bricks, difficult to transport (Wahida 1967: 172), may have been misidentified during the excavation according to J.D. Forest who identified half-bricks (Forest 1983: 13).

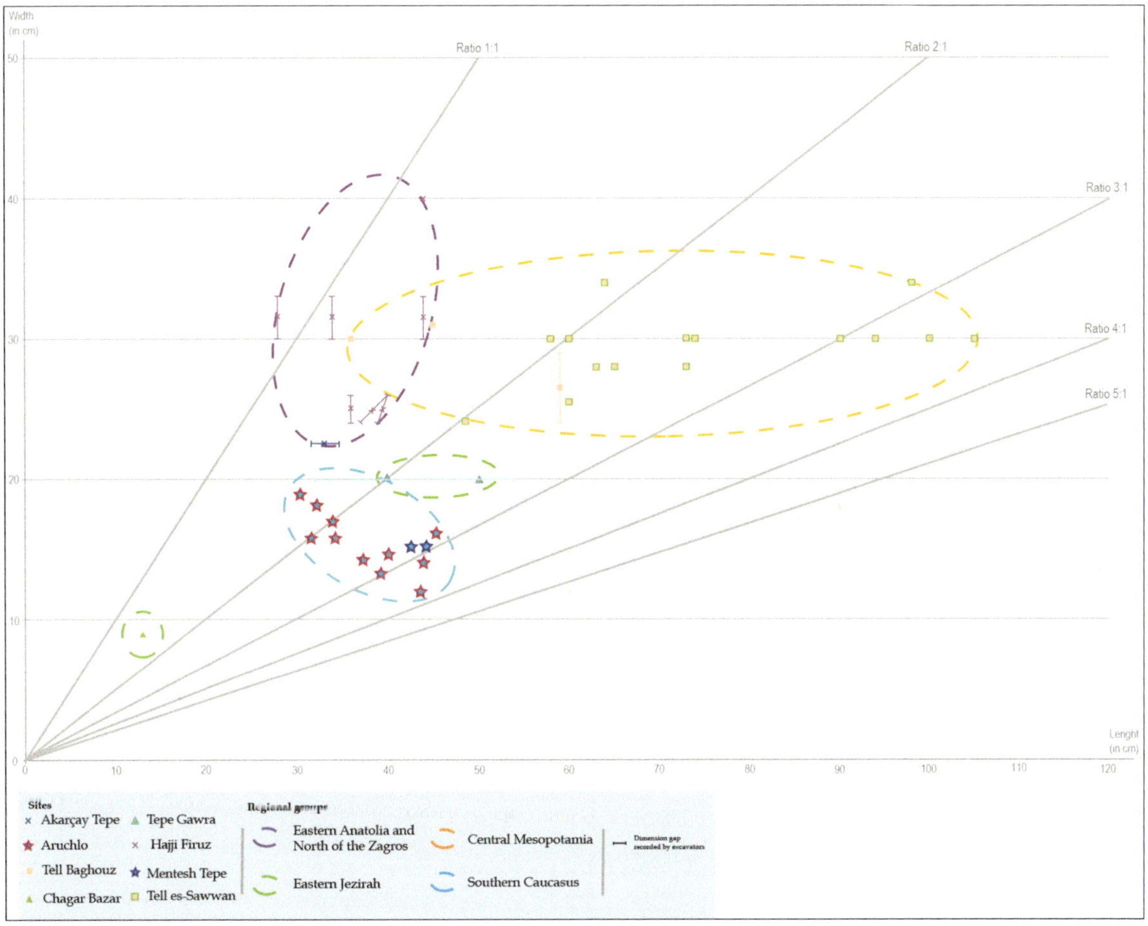


Figure 10. Summary diagram of the attestations of the molded mudbrick in Northern Mesopotamia and Caucasus (seventh-sixth millennium BCE)

In the Hamrin Basin, mudbricks at Choga Mami n°45-46 (Late Samarra, level III), are elongated (ratio: 3,3 to 7,5:1) and cigar-shaped like in the Halaf levels (level II), with fingerprints visible on the upper side (Oates 1969: 116, 117) and at Tell Sungur A n°187-188 during Late Samarra (Matsumoto 1984: 37; 1987: 189). In the Zagros foothills, similar mudbricks with fingerprints are produced at Choga Mish n°47-48 during Archaic Susiana 3 (Delougaz et al. 1972: 93) and at Choga Sefid n°49-51 (CMT) where the length reaches 160 to 260 cm (Hole 1977: 78). The maximum ratio of 1:7.3 by far exceeds the usual data, for an estimated weight of 93.6kg: such prefabricated elements are inconceivable to achieve because they would be difficult to transport without the risk of breaking the elements; it is more reasonable to interpret these 'bricks' as cob (Aurenche 1981: 195-199, but for an opposite argument, see Butterlin 2018: 124).

The molded mudbricks are attested before the middle of the seventh millennium BCE, in South-Eastern Anatolia, at Akarçay Tepe n°1-2 (Pottery Neolithic Phase, level F), where the excavators identified two different sizes with a ratio of 1.3 and 1.6:1 (Özbaşaran et al. 2011: 173).

After this period, in Northern Zagros, at Hajji Firuz n°81-88 (levels A to G), M. Voigt (1983: 19) identified molded mudbricks based on the steady of size and the regular shape close to square (Figure 10). In the Jezirah, at Chagar Bazar n°44 (Proto-Halaf, phase 10), the mudbricks used for circular buildings are shaped by hands or molded (Cruells et al. 2013: 471). At Yarim Tepe II n°203 (Halaf II, level VI), the wall of tholos 37 is made with molded mudbricks without, however, more precise information on their identification (Breniquet 1996: 84). At Tepe Gawra n°69 (Halaf, level XX), the regularity of size suggests that they are molded (Tobler 1950: 48).

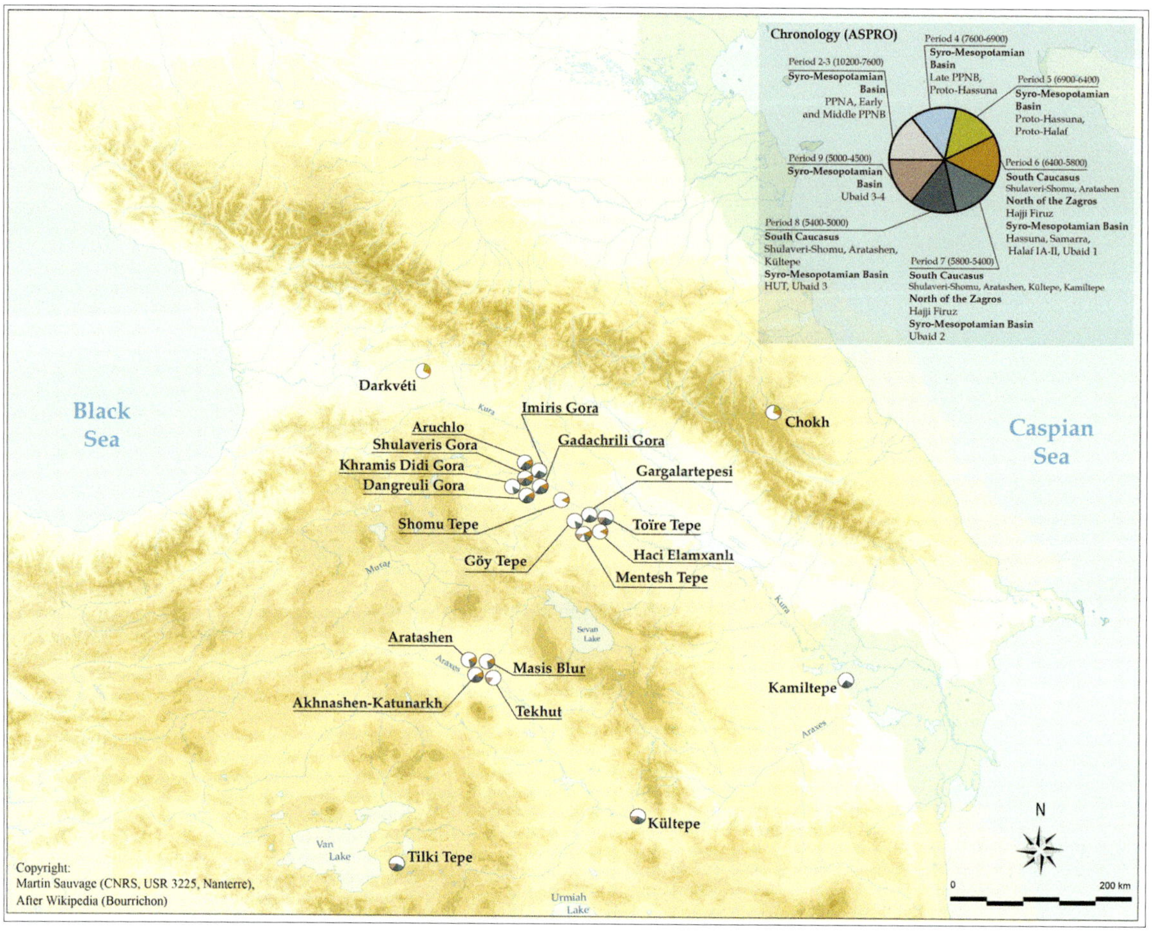


Figure 11. General map of Southern Caucasus with the sites mentioned in the text.

Figure 12. Gadachrili Gora, example of the wall 217 made of lumps (after Baudouin et al. 2018: 60, fig. 12).

In Central Mesopotamia, at Tell Baghouz n°36-38 (Late Samarra), R. Du Mesnil du Buisson (1948: 15; pl. XV–3) distinguished three different size and photographs enable to suppose that they are molded. At Tell es-Sawwan n°142-150, 153-164 (PN(?)/levels I-II), large oblong and molded mudbricks (El-Wailly et al. 1965: 21) are produced with a ratio between 2 and 3:1, while in level IV (Samarra) n°165-166, the molded mudbricks tend towards a ratio of 2:1 (Breniquet 1992: 9; El-Wailly et al. 1965: 21). The mudbricks at Tell es-Sawwan are considered by default as molded (Sauvage 1998: 101, 214). However, the weight of some bricks, reaching around 40 kg, makes them difficult both to make and transport.

### 3.2 Southern Caucasus communities

#### 3.2.1 Cob

In Southern Caucasus, recent discoveries record the use of lumps (Figures 4 and 11) in the Kvemo-Kartli (Georgia) and in the Ararat (Armenia) plains. At Gadachrili Gora n°54 (Shulaveri-Shomu, horizon II), wall 217 is built with lumps (Figure 12) (Baudouin et al. 2018; Hamon et al. 2016: 165, fig. 23), like in walls 139 and 164 (Hamon et al. 2016: 164) and at Masis Blur n°113 (Aratashen, level I) (Hayrapetyan et al. 2014: 180).

In the Ararat Plain, layers of cob seem used at Aknashen-Katunarkh (Aratashen, level IV) for wall T5W4 where strips of yellow clay divided horizontally by strips of darker clay have been identified (C. Chataigner, personal communication). At Masis Blur n°114 (Aratashen, level I), wall S011 is made with 'straight row of alternating dark and light clay rectangles' (Hayrapetyan et al. 2014: 180).

Several occurrences of cob could not be defined in detail. In the Kvemo-Kartli Plain, at Aruchlo n°13-14 (Shulaveri-Shomu, level I), cob, which is there called *pisé* (Chataigner 1995: 59; Kushnareva et al. 1970: 22), has been used for buildings 4 to 9. At Shulaveris Gora n°188 (Shulaveri-Shomu), A. Sagona (1993: 456) described this technique as 'wattle-and-daub', while the excavators as *pisé* (Dzhavakhishvili et al. 1975: 203). In these two cases, stacked mud and plano-convex mudbricks are used simultaneously (see below). In Nakhichevan, at Kültepe n°112 (Kültepe), Abibullaev (1963: 157–58) mentioned *pisé*, without specifics.

#### 3.2.2 Mudbricks

In Southern Caucasus, two morphological types of sun-dried mudbricks have been identified: plano-convex mudbricks, flat on the inside and curved on the outside, characteristic of Shulaveri-Shomu in the Kura Valley, and elongated and flat mudbricks, characteristic of Aratashen in the Araxes Valley (Chataigner 1995: 64) but also used in the Middle Kura Valley and the Kvemo-Kartli Plain. The publications usually describe the shape without specifying the manufacturing technique. C. Chataigner is the only one to identify plano-convex bricks as mudbricks shaped by hands, similar to the curved bricks produced during the PPNA (9500-8700 BCE) in the Near East (Chataigner 1995: 57).

Plano-convex mudbricks seem to be used from the beginning of the sixth millennium BCE in Shulaveri-Shomu. At Shomu Tepe n°167-175 (Narimanov 1987: 86), they have a length/width ratio always between 2:1 and 3:1, and two sizes: small and large (Figures 7 and 8). At Mentesh Tepe n°118-123 (level I), sizes is relatively homogeneous, with a ratio between 2.7:1 and 3.6:1. While the mudbricks of structure ST346 are certainly shaped by hands, other could be mold-made (see below). At Göy Tepe n°70-72 (levels 1 to 14), plano-convex mudbricks in level 13 are replaced by flat mudbricks in the other levels (Guliyev et al. 2014: 5, Y. Nishiaki, personal communication); medium mudbricks have a length of 30 to 40cm, and large mudbricks are 40-60cm (Guliyev et al. 2009: 47; 2012: 74; 2014: 5). At Gargalar Tepesi n°61-68, there is a clear decrease in size from levels 1 to 3 (Narimanov 1992: 20–21).

In the Kvemo-Kartli Plain, at Aruchlo n°15-23, scholars insisted on the irregularity of the size, with length varying between 18 and 50 cm (Hansen et al. 2013: 390). In addition, the stratigraphic data enable to associate the largest mudbricks to the oldest levels (Hansen et al. 2017: 209, fig. 22; p. 210, fig. 23). At Shulaveris Gora n°176-186 (Dzhavakhishvili 1973: 19), the same irregularity is attested, length varying between 25 and 50cm (Kushnareva et al. 1970: 28), but no stratigraphic data enable to assume an evolution of the size. At Gadachrili Gora n°54-60 (level I), mudbricks of building 2004 are elongated and flat with a ratio of 2:1 (Baudouin et al. 2018). In building 2003, mudbricks of the peripheral wall have a plano-convex shape, with a ratio of 2:1, and those of the dividing wall are square and flat (Hamon et al. 2016: 161–62). At Khramis Didi Gora n°104-111, plano-convex mudbricks

Figure 13. Molded plano-convex mud-bricks from Mentesh Tepe. a.: Detail of the four courses of mud-bricks, wall 293 (Sector 10) – View from the West; b.: Detail of the mortar between the mud-bricks, the wall 293 – View from the West; c.: Detail of the first course of mud-brick, wall 285 (Sector 15) – View from the North-East (after Baudouin et al. 2018: 58, fig. 8).

have a ratio between 1.9 and 2.7:1. At Imiris Gora n°88-94, such mudbricks are of different size (Dzhaparidze et al. 1971: 28; Dzhavakhishvili 1973: 48), varying between 32 and 50cm, but the lack of stratigraphic data does not enable to interpret an evolution.

In the Ararat Plain (Aratashen), flat and elongated mudbricks are attested at Aratashen n°6 in level I (Badalyan et al. 2004: 402) and at Akhnashen-Katunarkh n°3-5 where small and large mudbricks are identified in level VII (Badalyan et al. 2014: 165). At Tekhut n°196 (levels II and XII), mudbricks are also flat with a ratio of 1.3:1 (Torosjan 1976: 23–27).

In the Mil Plain, at Kamiltepe n°101-103 (Kamiltepe, phase Kamiltepe I), if Narimanov (1992: 35) identified rectangular mudbricks, those used for the terrace, shaped by hand, are smaller and square (Helwing et al. 2017: 17).

Recent research at Aruchlo (Ioseliani 2017) and Mentesh Tepe (Baudouin et al. 2018) confirmed the use of molded plano-convex mudbricks during the first third of the sixth millennium BCE in Shulaveri-Shomu (Figure 7, Baudouin, tbp). At Mentesh Tepe n°110-111, mudbricks are clearly molded (Figure 13): the edges are perfectly straight and marks of clipping are visible at the top of these mudbricks (Baudouin 2018a: 150–51; Baudouin et al. 2017: 44-45, fig. 4). In addition, slight longitudinal ridges on the convex face of several mudbricks indicate a flattening of the surface before the removal of the frame. At Aruchlo n°24-35, molded plano-convex mudbricks have been identified in a burned collapse attributable to the oldest occupation. These mudbricks have a standard size and have specific stigmas (Ioseliani 2017: 282) similar of those observed at Mentesh Tepe.

## 4 Discussion

### *4.1 Technical evolution: the gradual replacement of cob by mudbricks*

Stacked mud is used from the end of the seventh to the middle of the sixth millennium BCE. Lumps seems to be older, attested from Hassuna (Proto-Hassuna), but we observe the recurrence of cob in layers in the Jezirah, Northern Zagros and Southern Caucasus since 6200-6000 BCE: its use could be simultaneous during Early Halaf/Halaf I (ca. 6010-5800 BCE), Hajji Firuz (ca. 6200-6000 BCE), Aratashen (ca. 6000-5750 BCE) and Shulaveri-Shomu (from 5900 BCE). The use of cob seems significant from the end of the seventh millennium to the first third of the sixth millennium BCE but it gradually disappears during the end of Hassuna and Halaf in the Syro-Mesopotamian Basin, replaced by mudbrick.

Until the middle of the sixth millennium BCE, and whatever the manufacturing technique, the production of mudbricks testifies to extremely marked regional facies throughout the region considered, identifiable by differences in shapes or sizes. The reason of this heterogeneity is probably related to cultural particularisms. Both in the Syro-Mesopotamian Basin (Halaf) and in the Southern Caucasus, the appearance of molded mudbricks is not well defined. In the first area, they seem to appear late and perhaps as a result of an Ubaid impulse (Breniquet 1996: 84), during the Halaf II period (ca. 5800-5400 BCE). Their discovery in the Proto-Halaf levels at Chagar Bazar and in the Hajji Firuz levels on the eponymous site suggests that the East of the Euphrates was able to benefit from a technical spread from the Middle Euphrates at the end of the seventh millennium (Sauvage 2009: 195). Indeed, the use of the molded mudbrick on several sites (Cafer Hoyük, Tell Halula, Bouqras, Gritille) of Late PPNB (ca. 7000 BCE) in this region encourages to follow this assumption. In Southern Caucasus, mudbricks are duly molded since the beginning of the sixth millennium BCE.

### *4.2 Cob development: technical transfer or convergence?*

The process of the spread of a technique can be considered through time and space. This process involves an imbalance between two (or more) communities (Leroi-Gourhan 1945: 460). However, other factors need to be considered, such as the technical degree of the communities and the natural setting. The former presence of cob in Jezirah and the large amount of archaeological evidence enable us to precisely track the technical evolution and its origin in Northern Zagros and in the Southern Caucasus communities.

#### *4.2.1 The development of Halaf architecture*

The assumption of the spread of cob in Halaf is conceivable. It was adopted gradually during the Halaf expansion, between 5900 and 5300 BCE, towards Central Mesopotamia, Western Jezirah and near the Van Lake in Northern Zagros (Figure 14): there would be a convergence between spatial organization and chronology, that can confirm the assumption of a slow spread, phenomenon already emphasized for ceramics (Breniquet 1996: 65).

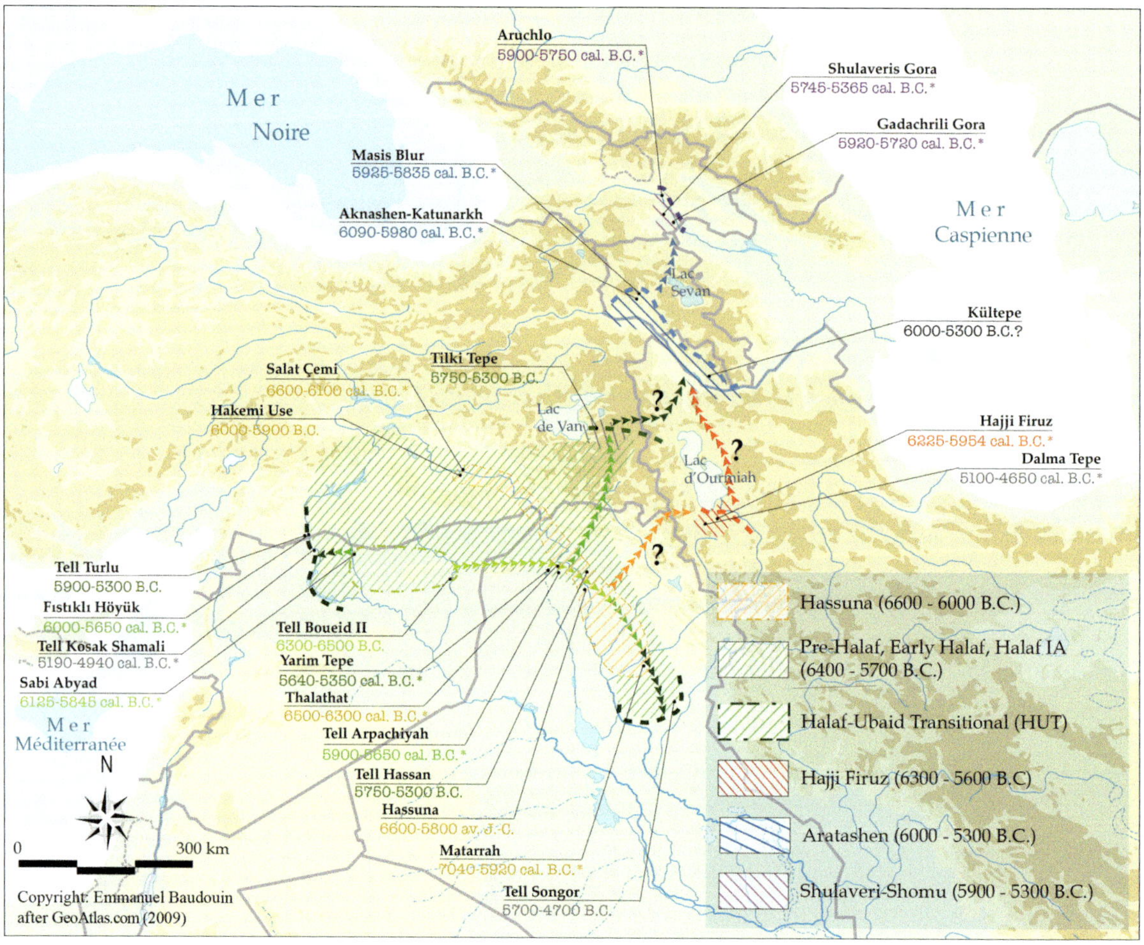


Figure 14. Use of cob and spread of the technique in the Syro-Mesopotamian basin, Eastern Anatolia, Northern Zagros and Southern Caucasus from the second half of the seventh millennium to the middle of the fifth millennium BCE.

### *4.2.2 The development of cob in Northern Zagros and Southern Caucasus*

An analysis of the stratigraphic sequence at Hajji Firuz brings new informations. Firstly, cob is used alone until levels G-F (Figure 15). Furthermore, this technique has been used in 'competition' with molded mudbrick from levels D to B. In this case, the molded mudbrick may be interpreted as a technical invention which will totally replace cob from levels C-B.

In the current state of knowledge, it is likely that cob appeared from the beginning of the Hajji Firuz (ca. 6300 cal. BCE), perhaps as result of the stimuli of the Eastern Jezirah or High Tigris communities during Early Hassuna or Samarra, especially because these cultures share extra-architectural affinities, as in decorations and ceramic forms (Voigt 1983: 163).

In the Southern Caucasus, the issue is different. Halaf reaches the shore of the Van Lake ca. 5700/5300 BCE, at least two hundred years after first settlement in Southern Caucasus. Therefore it seems unlikely that this technique was older than the Halaf expansion to the northeast, although there could have been relations between these different communities at the beginning of the sixth millennium BCE, as represented by the discovery of sherds of 'Mesopotamian' style at several sites. On the other hand, the possibility of a spread from the Urmiah Lake (Hajji Firuz) and High Tigris (Hassuna or Samarra) regions is weak because, except cob, technical affinities seem extremely thin with Caucasian architecture (Baudouin, in progress, 2019, 2018a 2018b). It would be unusual that the technical input only affected one specific material without changing other material components. Thus, in our current state of knowledge, a local and autonomous origin of cob in the Araxes Valley is likely. Nevertheless, an 'internal' spread in the Southern Caucasus from the Araxes

| LEVEL | RADIOCARBON DATING (CAL. B.C.) | COB (BUILDING N°) | MUD-BRICK (BUILDING N°) |
|---|---|---|---|
| L | 6225-5954 | NA | NA |
| K | / | NA | NA |
| J | / | XV2 | |
| G-H | / | XIII-2 | |
| F-G | / | XI1-2 | |
| E-F | / | | X2-3 |
| D | 5944-5592 | | IX |
| C | | VI1 | VI1 |
| C | | VI1 | |
| B-C | | | II2 |
| A3 | | | III |
| A3 | | | II1 |
| A2-A3 | | | I2 |

Figure 15. Summary table of the evolution of cob and mudbricks sizes at Hajji Firuz between levels L and A2, based on Voigt 1983.

to the Kura Valley is possible. However, lack of precise stratigraphic and chronological data does not enable for sure to support this assumption.

### *4.3. The mudbrick technology: a medium for a technical transfer?*

The interpretation of an autonomous invention of the molded mudbrick during Late PPNB (ca. 7500-7000 BCE) in the Middle Euphrates and a spread to Samarra in Central Mesopotamia (ca. 6500 BCE) is argued by Sauvage (2009: 195). In the Jezirah, the molded mudbrick would have been adopted late (Halaf II) and gradually at the impulse of Ubaid communities (Breniquet 1996: 95). Such a technical change would correlate with the social transformations within Halaf communities.

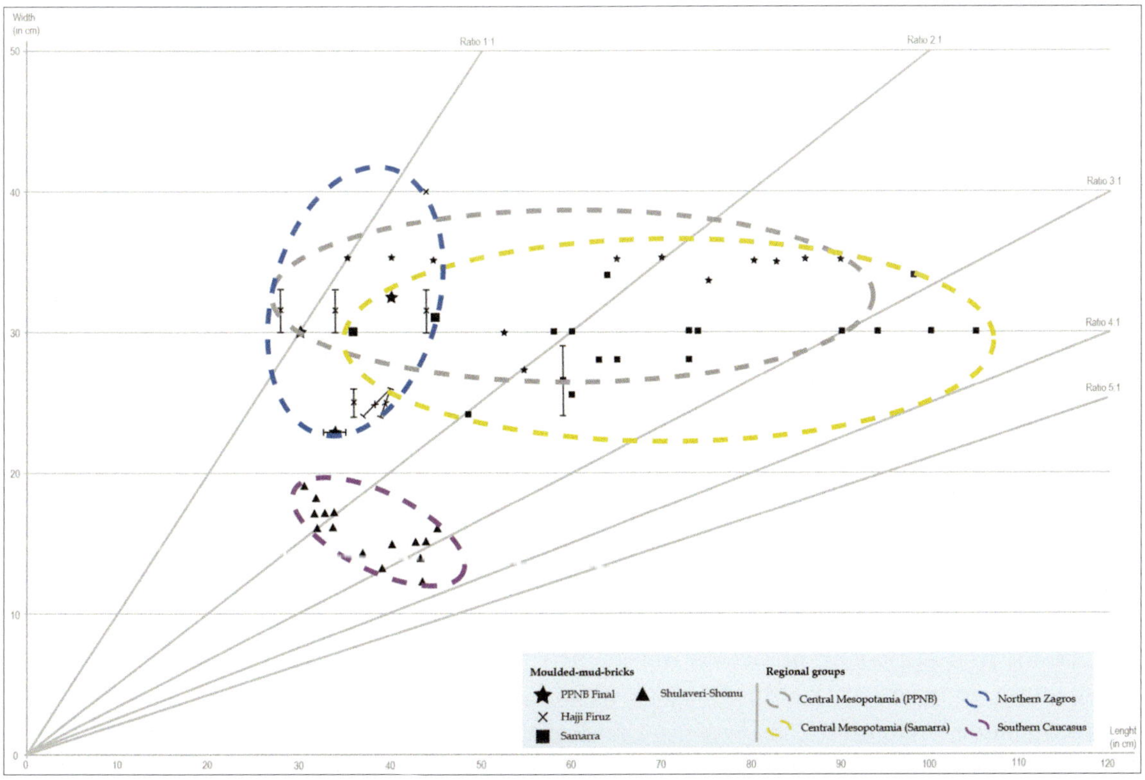


Figure 16. Sizes of molded mudbricks from the seventh millennium and the middle of the sixth millennium BCE in the Syro-mesopotamian basin, Northern Zagros and Southern Caucasus.

| LEVEL | DATE | BUILDING BUILT WITH COB TECHNIQUE | BUILDING BUILT WITH MUD BRICKS |
|---|---|---|---|
| 8 | 6125-6075 BC | 8.1 ; 8.4 | 8.2 ; 8.3 |
| 7 | 6020-5995 BC | 7.1 ; 7.2 ; 7.3 ; 7.5 | 7.5 ; 7.11 |
| 6 | 6010-5995 BC | 6.1 ; 6.2 ; 6.4 ; 6.5 ; 6.6 ; 6.7 ; 6.8 ; 6.9 ; 6.10 ; 6.11 | 6.1 ; 6.13 ; 6.14 |
| 5 | 6000-5945 BC | 5.1 | 5.1 ; 5.2 |
| 4 | 5980-5925 BC | / | Tholos Q15 |
| 3B | 5940-5905 BC | Build. I | Buil. I ; Tholos S ; Tholos I ; Tholos N/AE |

Figure 17. Summary table of the architectural techniques at Tell Sabi Abyad (Syria) with a gradual disappearance of the cob technique and its replacement by mudbricks, based on Akkermans et al. 2014 (after Baudouin et al. 2018: 67, fig. 17).

### *4.3.1 The development of mudbrick in Mesopotamia*

Chronologically, the assumption of a spread of the molded mudbrick from the Middle Euphrates to Central Mesopotamia is conceivable (Sauvage 2009: 195). It is explained by a 'late' occupation in the Syrian desert at El-Kowm 2, ca. 7900-5600 cal. BCE (Stordeur 2000: 305, tab.1), and Qdeir, ca. 7290-5730 cal. BCE (Hours et al. 1994: 408), or in the Euphrates Valley at Bouqras, ca. 7100-6200 cal. BCE (Hours et al. 1994: 388–89), which thus extends Late PPNB and reduces the gap with the beginning of Samarra, ca. 6400-5700 cal. BCE (El-Wailly et al. 1965: 19; Burleigh et al. 1982: 247; Helwing 2016). The similarity of the size of mudbricks in Late PPNB (Bouqras n°41-43, Gritille n°73-75) and Samarra (Figure 16) corroborates relationships between these two chrono-cultural areas and confirms the assumption of a spread of the molded mudbrick technique from the Middle Euphrates to Central Mesopotamia at the beginning of the sixth millennium BCE. In addition, this spread can be corroborated by other technical elements adopted in Central Mesopotamia like buttresses, attested for the first time at Cafer Höyük, during Late PPNB, ca. 8500-6500 cal. BCE (Molist et al. 1991: 7), or the standardized plan (Aurenche 1981: 200; Breniquet 2000; Stordeur 2000; Aurenche et al. 2009: 156; Baudouin 2018a: 497-508) attested at Tell Bouqras and at El-Kowm 2.

### *4.3.2 A technical improvement in Northern Zagros and Jezirah*

In Northern Zagros (Hajji Firuz), the typical mudbrick size at Hajji Firuz (Figure 16; in blue), with a ratio of 1:1, overlap very little with the production of Late PPNB (in grey) and Samarra (in yellow). The assumption of an autonomous technical evolution in Northern Zagros seems thus more viable and could explain the gradual disappearance of cob at the benefit of mudbrick (Figure 15). A similar assumption could be proposed in western Jezirah for hand-shaped mudbricks at Tell Sabi Abyad (Op. I) where we observed the same gradual displacement from level 8 to level 3B (6125-5905 cal. BCE), cob nearly totally disappearing ca. 6000 BCE (Figure 17).

### *4.3.3 How to interpret the original shape of the Caucasian mudbricks?*

Lastly, mudbricks made in the Kura Valley during the Shulaveri-Shomu are characteristic by their size and their shape (Figure 16). The hypothesis of a local invention would justify and corroborate the interpretation of an autonomous settling of communities in the Kura Valley. This assumption is strengthened by the presence of circular subterranean architecture in the Kura Valley, in the Mil Plain and in the Karabagh Plain at where in few sites these structures were built on the virgin soil and are the first settlement phases during the first third of the sixth millennium BCE This 'first' architecture, compared with ethnographic examples, can be considered as the original settlement of former nomadic communities (Baudouin 2019: 145-146). Finally, the exclusive use of the circular plan throughout more than 700 years in the entire Southern Caucasus enables us to interpret a technical inertia in the region and confirms the assumption of an autonomous settlement of communities of this region.

## 5 Conclusion

The spread of cob in the entire area of the Halaf cultural expansion is most likely. However, in Northern Zagros and Southern Caucasus, such a hypothesis is unlikely because no other technologies were adopted. The search the 'way' of spread of one single technique is constrained by chronological imprecisions and a lack of diagnostic markers. In addition, cob is often considered like a 'basic' technique characterizing communities who did not reach a high degree of technicality. Therefore, a local development in Northern Zagros and the Southern Caucasus, where the first settlement appeared at the beginning of the sixth millennium BCE, cannot be excluded. Later, the gradual disappearance of cob and its replacement by mudbrick is a phenomenon clearly visible at a local scale (Hajji Firuz, Tell Sabi Abyad), but it is not possible to generalize in the current state of data.

On the other hand, the appearance of molded mudbrick represents an undeniable technical development, the spread of which is clearly visible from Late PPNB to Samarra, and can only be approached by a general study of the 'complex' architecture in the Syro-Mesopotamian Basin. Therefore, the development of the Ubaid tripartite architecture represents a new stage of progress at the end of the fifth millennium BCE (Baudouin 2018a; 2018b: 407). This brief overview of the development of cob and mudbrick techniques shows the potential of this research and confirms the importance of systematizing such an approach in order to better define the contents of technical exchanges and cultural relationships between communities.

## Bibliography


Abibullaev, O. (1959) *Arkheeologicheskie Rakopki v Kyultepe* [Archaeological Excavations at Kültepe]. Baku: AN Azerbajdžanskoj SSR. Sovetskaja Arkheologija.

Abibullaev, O. (1963) 'Nekotorye Itogi Izuchenija Kholma Kjul Tepe v Azerbajdzhane [Elements for the Study of Kül Tepe. Archaeological Researches in Azerbaijan]'. *Sovetskaja Arkheologija*, 3: 157–68.

Abibullaev, O. (1982) *Eneolit i Bronza Na Territorii Nakhichevanskoj ASSR* [Eneolithic and Bronze Age in the Territory of Nakhichevan ASSR]. Baku: Elm.

Akkermans, P. (2000) 'Old and New Perspectives on the Origins of the Halaf Culture' in Rouault, O., Wäfler, M. (eds) *La Djéziré et l'Euphrate syriens de la protohistoire à la fin du IIe millénaire av. J.-C. : Tendances dans l'interprétation historique des données nouvelles*. Turnhout: Brepols, Subartu, VII: 43-54.

Akkermans, P. (2010) 'Late Neolithic Architectural Renewal: The Emergence of Round Houses in the Northern Levant, c. 6500-6000 BC.' in Bolger, D., Maguire, L.C. (eds) *The Development of Pre-State Communities in the Ancient Near East*. Oxford, Oxbow: 22-28.

Akkermans, P., Brüning, M., Huigens, H. and Nieuwenhuyse, O. (2014) *Excavations at Late Neolithic Tell Sabi Abyad, Syria: The 1994-1999 Field Seasons*. Turnhout: Brepols.

Aliyev, T., Helwing, B. (2009) 'Kamiltepe in der Milebene. Archaölogische Untersuchungen 2009', *Archäologische Mitteilungen aus Iran und Turan*, 41: 23-45.

Arimura, M., Badalyan, R., Gasparyan, B. and Chataigner, C. (2010) 'Current Neolithic Research in Armenia', *Neo-Lithics*, 1(10): 77–85.

Aurenche, O. (1977) (ed.) *Dictionnaire illustré multilingue de l'architecture du Proche-Orient ancien*. Lyon/Paris: TMO/De Boccard.

Aurenche, O. (1981) *La Maison Orientale : L'architecture Du Proche-Orient Ancien Des Origines Au Milieu Du Quatrième Millénaire*, Bibliothèque Archéologique et Historique, 109. Paris: P. Geuthner.

Aurenche, O. (1993) 'L'origine de la brique dans le Proche Orient ancien' in Frangipane, M., Hauptmann, H., Liverani, M., Matthiae: and Mellink, M. (eds) *Between the Rivers and Over the Mountains: Archaeologica Anatolica et Mesopotamica Alba Palmieri Dedicata*. Roma: Dipartimiento di Scienze Storiche Archeologiche e Anthropologiche dell'Antichità, Universtà di Roma La Sapienza: 71–86

Aurenche, O. and Calley, S. (1988) 'L'architecture de l'Anatolie du sud-est au Néolithique', *Anatolica*, XV1: 1-24.

Aurenche, O., Calley, S., Cauvin, J. and Molist, M. (1985) 'L'architecture de Cafer Höyük (fouilles 1982-1983): rapport préliminaire', *Cahiers de l'Euphrate*, 4: 11–33.

Aurenche, O., Klein, A. and Chazelles, C.A. (2011) 'Essai de classification des modalités de mise en œuvre de la terre crue en parois verticales et de leur nomenclature' in Chazelles, C.A., Klein, A., Pousthomis-Dalle, N. (eds) *Les cultures constructives de la brique crue: troisièmes échanges transdisciplinaires sur les constructions en terre, placés sous la présidence du professeur Olivier Aurenche: actes du colloque international de Toulouse, 16 et 17 mai 2008.* Montpellier: Éd. de l'Espérou: 13-34.

Aurenche, O. and Kozlowski, S.K. (2009) *La Naissance du Néolithique au Proche-Orient.* Paris: Éd. du CNRS.

Azimov, M.S. (2006) *Architectural Development in the Earliest Settled Agricultural Phases of Azerbaijan.* Oxford: BAR International Series 1467.

Badalyan, R. and Harutyunyan, H. (2014) 'Aknashen – The Late Neolithic Settlement of the Ararat Valley: Main Results and Prospects for the Research' in Gasparyan, B., Arimura, M. (eds) *Stone Age of Armenia. A Guide-book to the Stone Age Archeaology in the Republic of Armenia.* Kanazawa: Kanazawa University, Center for Cultural Resource Studies: 161-76.

Badalyan, R., Lombard, P., Avetisyan, P., Chataigner, C., Chabot, J., Vila, E., Hovespyan, R., Willcox, G., Pessin, H. and Lyonnet, B. (2007) 'New Data on the Late Prehistory of the Southern Caucasus. The Excavations at Aratashen (Armenia): Preliminary Report' in Lyonnet, B. (ed.) *Les Cultures Du Caucase (VIe-IIIe Millénaires Avant Notre Ère: Leurs Relations Avec Le Proche-Orient.* Paris: CNRS Éditions: 37-62.

Badalyan, R., Lombard, P., Avetisyan, P. and Chataigner, C. (2004) 'The Neolithic and Chalcolithic Phases in the Ararat Plan (Armenia): The view from Aratashen' in Sagona, A. (ed.) *A view from the highlands: archaeological studies in honour of Charles Burney.* Leuven: Peeters: 399–420.

Baudouin, E. (2017) 'Remarques préliminaires sur les techniques architecturales du Néolithique et du Chalcolithique à Mentesh Tepe (Azerbaijan)' in Helwing, B., Aliyev, T., Lyonnet, B., Guliyev, F., Hansen, S. and Mirtskhulava, G. (eds) *The Kura Projects: New Research on the Late Prehistory of the Southern Caucasus.* Berlin: Deutsches Archäologisches Institut, Eurasien-Abteilung, Archäologie in Iran und Turan, 16: 153-62.

Baudouin, E. (2018a) 'L'architecture en Syro-Mésopotamie et dans le Caucase de la fin du 7e à la fin du 5e millénaire' Thèse de doctorat. Paris: Sorbonne Université.

Baudouin, E. (2018b) 'L'architecture en Syro-Mésopotamie et dans le Caucase de la fin du VIIe à la fin du Ve Millénaire', *Bulletin de La Société Préhistorique Française*, 115 (2): 405–407.

Baudouin, E. (2019) 'Rethinking architectural techniques of the Southern Caucasus in the 6th millennium BC: A re-examination of former data and new insights', *Paléorient*, 45 (1): 115–150.

Baudouin, E. (2020) 'Modelées ou moulées ? Les premières briques crues du Caucase au Néolithique', e-Phaïstos, VIII (2). Available at: https://journals.openedition.org/ephaistos/7912 [Accessed January 30, 2021]

Baudouin, E. (tbp) L'architecture en Syro-Mésopotamie et dans le Caucase de la fin du 7e à la fin du 5e millénaire. Turnhout: Brepols, ARAXES.

Baudouin, E., Lyonnet, B., Hamon, C. (2018) 'Architectural Techniques and Cultural Relationships between the Caucasus and Mesopotamia at the Beginning of the Sixth Millennium B.C.' in Douché, C. and Pichon, F. (eds) *Routes de l'Orient, Actes I, From the Caucasus to the Arabien Peninsula: Domestic Spaces in the Neolithic,* Routes de l'Orient: 49-84. accessed July 21 2019: https://rdorient.hypotheses.org/948

Baudouin, E., Mura, M. and Manel, C. (2017) 'Mentesh Tepe (Azerbaïdjan) : Résultats préliminaires de la campagne de fouille 2015', *Routes de l'Orient*, 3: 41-52.

Benecke, N. (2017) 'Exploitation of Animal Resources in Neolithic Settlements of the Kura Region (South Caucasia)'in Helwing, B., Aliyev, T., Lyonnet, B., Guliyev, F., Hansen, S. and Mirtskhulava, G. (eds) *The Kura Projects: New Research on the Late Prehistory of the Southern Caucasus.* Berlin: Deutsches Archäologisches Institut, Eurasien-Abteilung, Archäologie in Iran Und Turan, 16: 357-369.

Bernbeck, R., Pollock, S., Allen, S., Castro, A.G., Keilt, S., Costello ,R., Foree, M., Gleva, M., Goodwin, M., Lepinski, S., Nakamura, C. and Niebuhr, S. (2003) 'The Biography of an Early Halaf Village: Fıstıklı Höyük 1999-2000', *Istanbuler Mitteilungen*, 53: 9-77.

Braidwood, R.J., Braidwood, L., Smith, J.G. and Leslie, C. (1952) 'Mattarah: A Southern Variant of the Hassunan Assemblage, Excavated in 1948', *Journal of the Near Eastern*, 11: 2–75.

Breniquet, C. (1987) 'Note sur les principaux résultats de la fouille de Tell Turlu 1962', *Paléorient*, 13(1): 113-116.
Breniquet, C. (1992) 'Rapport sur deux campagnes de fouilles à Tell es-Sawwan 1988-1989', *Mesopotamia*, 27: 5-30.
Breniquet, C. (1996) *La disparition de la culture de Halaf: les origines de la culture d'Obeid dans le Nord de la Mésopotamie*. Paris: Éd. Recherche sur les civilisations.
Breniquet, C. (2000) 'De Bouqras à Tell es-Sawwan. Les premiers avatars des plans tripartites du nord mésopotamien' in Rouault, O. and Wäfler, M. (eds) *La Djéziré et l'Euphrate Syriens de la Protohistoire à la fin du IIe millénaire av. J.-C. : Tendances dans l'interprétation historique des données nouvelles*. Turnhout: Brepols, Subartu, VII: 55-70.
Breniquet, C. (2016) 'Recension de Nieuwenhuyse, O.P., Bernbeck, R., Akkermans, P.M.M.G and Rogasch, J. (eds) 2013. 'Interpreting the Late Neolithic of Upper Mesopotamia', Turnhout: Brepols (Papers on Archaeology of the Leiden Museum of Antiquities, Palma 9). 520 P', *Paléorient*, 42 (1): 207-209.
Burleigh, R., Matthews, K. and Ambers, J. (1982) 'British Museum Natural Radiocarbon Measurements XIV', *Radiocarbon*, 24(3): 229-61.
Butterlin, P. (2009) 'Introduction: Tepe Gawra et le monde proto-urbain de Mésopotamie' in Butterlin: (ed.) *À propos de Tepe Gawra, le monde proto-urbain de Mésopotamie. About Tepe Gawra: a proto-urbain world in Mesopotamia*. Turnhout: Brepols, Subartu, XXIII: 1-14.
Butterlin, P. (2018) *Architecture et Société Au Proche-Orient Ancien, Les Manuels d'arts et d'archéologie Antiques*. Paris: Picard.
Campbell, S. (2007) 'Rethinking Halaf Chronologies', *Paléorient*, 33(1): 103-36.
Chataigner, C. (1995) *La Transcaucasie au Néolithique et au Chalcolithique*. Oxford: Tempus Reparatum.
Chataigner, C., Badalyan, R. and Arimura, M. (2014) 'The Neolithic of the Caucasus', accessed July 21 2019: https://doi.org/10.1093/oxfordhb/9780199935413.013.13.
Chazelles C.A. de (1999) 'Les constructions en terre de la Gaule méditerranéenne' in Hammam, M. (ed.) *L'architecture de terre en Méditerranée, actes du colloque de Rabat (Maroc), 27-29 Novembre 1996*, Série Colloques et Séminaires. Rabat: Université Mohamed V: 15-40.
Chazelles-Gazzal, C.A. (1997) *Les Maisons en terre de la Gaule méridionale*. Montagnac: Monique Mergoil.
Chazelles C.A. de and Klein, A., eds. (2003) *Échanges transdisciplinaires sur les constructions en terre: actes de la table-ronde de Montpellier 17-18 Novembre 2001. 1, Terre modelée, découpée ou coffrée: matériaux et modes de mise en œuvre*. Montpellier: Éd. de l'Espérou.
Chazelles C.A. de, Klein, A. and Pousthomis-Dalle, N., eds. (2011) *Les cultures constructives de la brique crue: troisièmes échanges transdisciplinaires sur les constructions en terre, placés sous la présidence du professeur Olivier Aurenche: actes du colloque international de Toulouse, 16 et 17 Mai 2008*. Montpellier: Éd. de l'Espérou.
Chubinshvili, T.N., Niebieridze, L.D. and Pkhakadze, G.G. (1976) 'Pamyatniki rannezemledelcheskoy kultury v zone stroyitelstva Arukhlinskoy orositelnoy sistemy [Sites of the early land-tilling culture in the construction zone of the Arukhlo irrigation system]', *Arkheologischeskiye issledovaniya na novostroykakh Gruzinskoy SSR*: 55-62.
Contenson, H. de (1985) 'La Campagne de 1965 à Bouqras', *Cahiers de l'Euphrate*, 4: 335-371.
Cruells, W., Gomes, A., Bouso, M., Guerrero, E., Tornero, C., Sana, M., Molist, M., Buxo, R., Badhdo, A. and Tunca, Ö. (2013) 'Chagar Bazar in Northeastern Syria: Recent Work' in Bernbeck, R., Akkermans, P., Rogasch, J. and Nieuwenhuyse, O., *Interpreting the Late Neolithic of Upper Mesopotamia*. Turnhout: Brepols. pp. 467-478.
Cruells, W. and Nieuwenhuyse, O. (2003) 'The Proto-Halaf Period in Syria. New Sites, New Data', *Paléorient*, 30(1), pp.47-68.
Delougaz P. and Kantor, H.J. (1972) 'The 1971 Season of Excavation of the Joint Iranian Expedition', *Bastan Chenassi va Honar-e-Iran*, 9/10: 88-96.
Doat, P., Hays, A., Houben, H., Matuk, S. and Vitoux, F. (1979) *Construire en terre*. Grenoble: Alternatives et parallèles.
Dzhaparidze, O. and Dzhavakhishvili, I. (1971) 'Rezul'taty Rabot Kvemo-Kartlijskoj Arkheologicheskoj Ekspedisii (1968-1969 Gg) [Excavation Results of the Archaeological Expedition in the Kvemo-Kartli Plain (Campaigns 1968-1969)]', *Otchety Arkheologischeskikh Ekspeditisii Gos. Muzeia Gruzzi AN GSSR*, 2: 22-33.

Dzhavakhishvili, A.I. (1973) *Stroitel'noe Delo I Arkhitektura Poseleij Juzhnogo Kavkaza V-III Tys. Do n.e* [Architecture on the Settlements in South Caucasus V-III Mill. BC]. Tbilissi: Metsniereba.

Dzhavakhishvili, A.I. and Dzhaparidze, O.M. (1975) *Otchet Kvemo-Kartlijskoj arkheologicheskoj ekspeditsii (1965-1971 gg)* [Report on Kvemo-Kartili Archeaological Expedition (1965/1971)]. Tbilissi: Metsniereba.

El-Wailly, F. and Abu es-Soof, B. (1965) 'The Excavations at Tell Es-Sawwan. First Preliminary Report (1964)', *Sumer*, 21: 17-32.

Forest, J.D. (1983) 'Aux origines de l'architecture obeidienne: les plans de type Samarra', *Akkadica*, 34: 1-47.

Forest, J.D. (1996) *Mésopotamie: l'apparition de l'Etat - VIIe-IIIe millénaires.* Paris: Méditerranée.

Fuji, H. (1981) 'Preliminary Report of Excavations at Gubba and Songor', *Al-Rafidan*, 2: 2-342.

Fukai, S., Horiuchi, K. and Matsutani, T. (1970) *Telul eth-Thalathat. 2, The excavation of tell II. The Third season (1964).* Tokyo: Institute of Oriental Culture, University of Tokyo.

Fukai, S. and Matsutani T. (1981) *Telul eth-Thalathat IV. The Excavation of Tell II, Fifth Season (1976).* Tokyo: The Institute of Oriental Culture, University of Tokyo.

Guillaud, H., Chazelles, C.A. de and Klein, A. eds. (2007) *Les Constructions en terre massive: pisé et bauge - deuxièmes échanges transdisciplinaires sur les constructions en terre crue: actes de la table-ronde de Villefontaine, Isère, 28-29 Mai 2005.* Montpellier: Éd. de l'Espérou.

Guliyev, F. and Nishiaki, Y. (2012) 'Excavations at the Neolithic Settlement of Göytepe, the Middle Kura Valley, Azerbaijan 2008-2009' in Matthews, R. and Curtis, J. (eds) *Proceedings of the 7th International Congress on the Archaeology of the Ancient Near East: 12 April-16 April 2010, the British Museum and UCL, London - Volume 3 : Fieldwork & Recent Resarch Posters.* Wiesbaden: Harrassowitz Verlag: 71-84.

Guliyev, F. and Nishiaki, Y. (2014) 'Excavations at the Neolithic Settlement of Göytepe, West Azerbajan 2010–2011'in Bielinski:, Gawlikowski, M., Kolinsk, R., Lawecka, D., Soltysiak, A., and Wygnanska, Z. (eds) *Proceedings of the 8th International Congress on the Archaeology of the Ancient Near East. 30 April–4 May 2012, University of Warsaw - Volume 2: Excavation and Progress Reports Posters.* Wiesbaden: Harrassowitz Verlag: 3-16.

Guliyev, F., Nishiaki, Y., Hüseynov, F., Kadowaki, S., Tanno, K. Hayakawa, Y., Hacizade, N., Babayeva, T., Kume, S., Shimogama, K. Arimatsu, Y. and Akashi, C. (2009) 'Göytepe Neolit Yaşayiş Yerinde Arkeoloji Qazintilar (Tovuz Rayonu)'in *Azerbaycanda Arxeoloji Tedqiqatlar.* Baku: Arxeologiya ve Etnoqrafiya Institutu: 45-54.

Hamlin, C. (1975) 'Dalma Tepe', *Iran*, 13: 111-127.

Hamon C., Jalabadze M., Agapishvili T., Baudouin E., Koridze I. and Messager E. (2016) 'Gadachrili Gora: A New Neolithic Excavation within the Shulaveri Group (Chramis Valley, Georgia)', *Quaternary International*, 395: 154-169.

Hansen, S., Bastert-Lambrichs, K., Kromer, B. and Ullrich, M. (2017) 'Stratigraphy and Radiocarbon Dating' in Helwing, B., Aliyev, T., Lyonnet, B., Guliyev, F., Hansen, S., and Mirtskhulava, G. (eds) *The Kura Projects: New Research on the Late Prehistory of the Southern Caucasus.* Berlin: Deutsches Archäologisches Institut, Eurasien-Abteilung, Archäologie in Iran und Turan, 16: 291-96

Hansen, S. and Mirtskhulava, G. (2012) 'The Neolithic Settlement of Aruchlo. Report on the Excavations in 2009-2011', in Lyonnet, B., Guliyev, F., Helwing, B., Aliyev, T., Hansen, S., and Mirtskhulava, G. (eds), Ancient Kura 2010-2011: The First Two Seasons of Joint Field Work in the Southern Caucasus. *Archäologische Mitteilungen Aus Iran Und Turan*, 44: 58-70.

Hansen, S., Mirtskhulava, G. and Bastert-Lambrichs, K. (2013) 'Neolithic Settlements of the 6th Millennium cal. BCE in the Southern Caucasus' in Nieuwenhuyse. O., Bernbeck, R., Akkermans, P. and Rogasch, J., *Interpreting the Late Neolithic of Upper Mesopotamia.* Turnhout: Brepols: 387-396.

Hansen, S., Mirtskhulava, G., Bastert-Lambrichs, K., Benecke, N., Gatsov, I., and Nedelcheva, P. (2006) 'Aruchlo 2005-2006 Bericht Über Die Ausgrabungen in Einem Neolithischen Siedlungshügel', *Archäologische Mitteilungen Aus Iran Und Turan*, 38: 1-34.

Hansen, S., Mirtskhulava, G., Bastert-Lambrichs, K., Görsorf, J., Neumann, D., Ullrich, M., Gatsov, I., and Nedelcheva, P. (2007) 'Aruchlo 2007 Bericht Über Die Ausgrabungen in Einem Neolithischen Siedlungsügel', *Archäologische Mitteilungen Aus Iran Und Turan*, 39: 1-30.

Harutyunyan, A. (2014) 'Aknashen – On the Neolithic Pottery from the Settlement of Aknashen in the Ararat Valley' in Gasparyan, B. and Arimura, M. (eds), *Stone Age of Armenia. A Guide-Book to the Stone Age Archeaology in the Republic of Armenia.* Kanazawa: Center for Cultural Resource Studies: 191-204.

Hayrapetyan, A., Martyrosyan-Olshansky, K., Areshian, G.E. and Avetisyan, P. (2014) 'Preliminary Results of the 2012 Excavations at the Late Neolithic Settlement of Masis Blur' in Gasparyan, B. and Arimura, M. (eds) *Stone Age of Armenia. A Guide-Book to the Stone Age Archeaology in the Republic of Armenia.* Kanazawa: Center for Cultural Resource Studies: 177-90.

Helwing, B. (2016) 'Reconsidering the Neolithic Graveyard at Tell es-Sawwan, Iraq'. *Paléorient*, 42(1): 129-142.

Helwing, B. and Aliyev, T. (2012) 'Field Work in the Mil Plain: The 2010-2011 Expedition' in Lyonnet, B., Guliyev, F., Helwing, B., Aliyev, T., Hansen, S. and Mirtskhulava, G. (eds) Ancient Kura 2010-2012: The First Two Seasons of Joint Field Work in the Southern Caucasus. *Archäologische Mitteilungen Aus Iran Und Turan*, 44: 4-17.

Helwing, B. and Aliyev, T. (2017) 'Excavations in the Mil Plain Sites 2012-2014'in Helwing, B., Aliyev, T., Lyonnet, B., Guliyev, F., Hansen, S. and Mirtskhulava, G. (eds) *The Kura Projects: New Research on the Late Prehistory of the Southern Caucasus.* Berlin: Deutsches Archäologisches Institut, Eurasien-Abteilung, Archäologie in Iran Und Turan, 16: 11-42.

Hole, F. (1977) *Studies in the Archaeological History of the Deh Luran Plain: The Excavation of Chagha Sefid.* Ann Arbor: Memoirs of the Museum of Anthropology, University of Michigan.

Hours, F., Aurenche, O., Cauvin, J. and Lombard: (1994) *Atlas des sites du Proche Orient (14000-5700 BP).* Lyon/Paris: Maison de l'Orient méditerranéen/De Boccard.

Ioseliani, V. (2017) 'Architecture Debris from the Burnt Layers' in Helwing, B., Aliyev, T., Lyonnet, B., Guliyev, F., Hansen, S. and Mirtskhulava, G. (eds) *The Kura Projects: New Research on the Late Prehistory of the Southern Caucasus.* Berlin: Deutsches Archäologisches Institut, Eurasien-Abteilung, Archäologie in Iran und Turan, 16: 281-286.

Ippolitoni, F. (1970) 'The Pottery of Tell es-Sawwan', *Mesopotamia*, 5-6: 105-179.

Jasim, S. (1985) *The Ubaid Period in Iraq: The Recent Excavations in the Hamrin Region.* Oxford: British Archaeological Reports 267.

Kantor, H. (1978) 'Choga Mish and Choga Banut', *Iran*, 16: 189-191.

Kenyon, K.M. (1957) *Digging up Jericho.* London: Ernest Benn Limited.

Kiguradze, T. (1976) *Aghmosavlet Amierkavkasiis Adresamitsatmokmedo Lulturis Periodiatsia.* Tbilissi, Mestniereba.

Kiguradze, T. (1986) *Neolithische Siedlungen von Kvemo-Kartili, Georgien.* München: Beck, Materialien Zur Allgemeinen Und Vergleichenden Archäologie.

Kirkbride, D. (1972) 'Umm Dabaghiyah 1971: A Premilinary Report and Early Ceramic Farming Settlement in Marginal North Central Jazira. Iraq', *Iraq*, 34: 3-17.

Kozlowski, S.K. (1989) 'Nemrik 9, a PPN Neolithic Site in Northern Iraq', *Paléorient*, 15(1): 25-31.

Kozlowski, S.K. and Kempisty, A. (1990) 'Architecture of the Pre-Pottery Neolithic Settlement in Nemrik', *World Archaeology*, 21(3): 348-362.

Kushnareva, K.K. (1997) *The Southern Caucasus in Prehistory: Stages of Cultural and Socioeconomic Development from the Eighth to the Second Millennium BC.* Philadelphia: University Museum, University of Pennsylvania.

Kushnareva, K.K. and Chubinshvili, T.N. (1970) *Drevnie Kul'tury Ûžnogo Kavkaza (V-III Tys. Do n. è.)* [Old Cultures of Southern Caucasus]. Leningrad: Nauka.

Lawn, B. (1974) 'University of Pennsylvania Radiocarbon Dates XVII', *Radiocarbon*, 16(2): 219-237.

Le Mière, M. (2000) 'L'occupation Proto-Hassuna du Haut-Khabur occidental d'après la céramique' in Lyonnet, B. (ed.) *Prospection archéologique du Haut-Khabur occidental (Syrie du Nord-Est) - Volume I.* Beyrouth: Institut Français d'Archéologie du Proche-Orient: Geuthner, Bibliothèque Archéologique et Historique, 155: 127-150.

Leroi-Gourhan, A. (1945) *Evolution et techniques [2], Milieu et techniques.* Paris: Albin Michel.

Leroi-Gourhan, A. (1988) *Dictionnaire de la préhistoire.* Paris: PUF, Quadrige.

Lisitsyna, G.N. and Prischpenko, L.V. (1977) *Paleoetnobotanicheskie Nakhodki Kavkaa i Blizhnego Vostoka* [Paleobotanical Data from the Caucasus and the Near East]. Moscou: Nauka.

Lloyd, S. and Safar, F. (1945) 'Tell Hassuna: Excavations by the Irak Government Directorate of Antiquities in 1943 and 1944', *Journal of Near Eastern Studies*, 4: 255-289.

Lombard, P. and Chataigner, C. (2004) 'Le Néolithique et le Chalcolithique en Transcaucasie: l'exemple des bassins de la Kura et de l'Araxe' in Guilaine, J. (ed.), *Aux marges des grands foyers du Néolithique.* Paris: Errance: 61-84.

Lyonnet, B. and Guliyev, F. (2012) 'Recent Research on the Chalcolithic Period in Western Azerbaijan' in Matthews, R. and Curtis, J. (eds) *Proceedings of the 7th ICAANE, London 2010.* Wiesbaden: Harrassowitz: 85-98.

Lyonnet, B., Guliyev, F., Baudouin, E., Bouquet, L., Bruley-Chabot, G., Samzun, A., Fontugne, M., Degorre, E., Husson, X. and Raymond, P. (2017) 'Mentesh Tepe (Azerbaijan), a Preliminary Report on the 2012-2014 Excavations' in Helwing, B., Aliyev, T., Lyonnet, B., Guliyev, F., Hansen, S., and Mirtskhulava, G. (eds) *The Kura Projects: New Research on the Late Prehistory of the Southern Caucasus.* Berlin: Deutsches Archäologisches Institut, Eurasien-Abteilung (Archäologie in Iran und Turan 16): 125-140.

Lyonnet, B., Guliyev, F., Bouquet, L., Bruley-Chabot, G., Samzun, A. Pecqueur, L., Jovenet, E., Baudouin, E., Raymond, P., Degorre, E., Astruc, L., Guilbeau, D., Le Dosseur, G., Benecke, N., Hamon, C., Poulmarc'h, M., and Courcier, A. (2016) 'Mentesh Tepe, an Early Settlement of the Shomu-Shulaveri Culture in Azerbaijan', *Quaternary International,* 395: 170-83.

Lyonnet, B., Guliyev, F., Helwing, B., Aliyev, T., Hansen, S. and Mirtskhulava, G. (eds) (2012) 'Ancient Kura 2010-2011: The First Two Seasons of Joint Field Work in the Southern Caucasus', *Archäologische Mitteilungen Aus Iran Und Turan,* 44: 1-191.

Mallowan, M. (1936) 'The Excavations at Tell Chagar Bazar', *Iraq,* 3: 1-86.

Mallowan, M. and Cruikshank, R. (1935) 'Excavations at Tall Arpachiyah 1933', *Iraq,* 2: 1-179.

Martyrosyan-Olshansky, K., Areshian, G.E., Avetisyan, P. and Hayrapetyan, A. (2013) 'Masis Blur: A Late Neolithic Settlement in the Plain of Ararat, Armenia', *Backdirt,* 12: 142-146.

Matsumoto, K. (1984) 'Tell Songor A, B and C in Hamrin', *Sumer,* 40: 37.

Matsumoto, K. (1987) 'The Samarra Period at Tell Songor A' in Huot, J.L. (ed.) *Préhistoire de la Mésopotamie: la Mésopotamie préhistorique et l'exploration récente du Djebel Hamrin.* Paris: CNRS: 189-98.

Matsutani, T. (1991) *Tell Kashkashok: the excavations at Tell no. II.* Tokyo: University of Tokyo Press.

Melgaard, J, Mortensen, P. and Thrane, H. (1963) 'Excavations at Tepe Guran, Luristan', *Acta Archaeologica,* 34: 97-133.

Mellaart, J. (1970) *Excavations at Hacilar.* Edinburgh: The University Press.

Merpert, N., Munchaev, R. and Bader, N. (1976) 'The Investigation of Soviet Expedition in Iraq 1973', *Sumer,* 32: 25-62.

Merpert, N., Munchaev, R. and Bader, N. (1977) 'The Investigation of Soviet Expedition in Iraq 1974', *Sumer,* 35(1): 65-104.

Mesnil du Buisson, R. de (1948) *Baghouz, l'ancienne Corsôtê: Le tell archaïque et la nécropole de l'âge du Bronze.* Leiden: E. J. Brill.

Miyake, Y. (2011) 'Salat Cami Yanı. A Pottery Neolithic Site in the Tigris Valley' in Özdoğan, M., Başgelen, N., and Kuniholm, P. (eds) *The Neolithic in Turkey: New Excavations & New Research - The Tigris Basin (Vol. 1).* Istanbul: Archaeology and Art Publications: 129-149.

Molist, M., Anfruns, J., Borrell, F., Clop X., Cruells, W., Gomes, A., Guerrero, E., Tornero, C., Sana, M. and Abdul Massih, J. (2007) 'Tell Halula (vallée de l'Euphrate, Syrie): nouvelles données sur les occupations néolithiques - notice préliminaire sur les travaux 2002-2004' in Abdul Massih, J. (ed.) *Les résultats du programme de formation à la sauvegarde du patrimoine culturel de Syrie. Cultural Heritage.* Damas: UE-DGAM: 4-19.

Molist, M. and Cauvin, J. (1991) 'Les niveaux inférieurs de Cafer Höyük (Malatya, Turquie): stratigraphie et architecture (fouilles 1984-1986)', *Cahiers de l'Euphrate,* 5–6: 85-114.

Molist, M., Gomez, A., Boffil, M., Cruells, W., Faura, J.M., Marchiori, C. and Martin, J. (2014) 'Maisons et constructions d'habitation dans le Néolithique. Une approche de l'évolution des unités d'habitat domestiques à partir des documents de Tell Halula (Vallée de l'Euphrate, Syrie)' in Montero Fenollos J.L. (ed.) *Redonner vie aux mésopotamiens: mélanges offerts à Jean-Claude Margueron à l'occasion de son 80e anniversaire.* Ferrol: Proyecto arqueológico Medio Éufrates Sirio, Sociedade Luso-Galega de Estudos Mesopotámicos: 107-26.

Mongne, P. and Marquis, P. (2008) *Dictionnaire de l'archéologie.* Paris: Larousse.

Munchaev, R. (1975) *Kavkazna Zare Bronzovogo Veke* [The Caucasus at the Dawn of the Bronze Age]. Moscou: Nauka.

Munchaev, R. (1982) 'Eneolit Kavkaza [The Eneolithic of the Caucasus]'in Masson, V., Merpert, N. and Chernysh, E., *Eneolit SSSR,* Moscou: Nauka: 93-164

Munchaev, R. and Merpert, N. (1973) 'Excavations at Yarim Tepe 1972 (Fourth Preliminary Report)', *Sumer,* 29: 3-16.

Munchaev, R. and Merpert, N. (1981) *Rannezemledel'cheskie poseleniya Severnoy Mesopotamii* [Earliest Agricultural Settlements of Northern Mesopotamia]. Moscou: Nauka.

Narimanov, I.G. (1965) 'Arkheologischeskie issledovanija poselenija Shomu Tepe v 1963 G. [The archeological investigations of Shomu Tepe settlement in 1963]. Arkheologicheskie Issledovanija v Azerbajdzhane, pp.45-53.

Narimanov, I.G. (1986) 'Raskopki Eneoliticheskikh Poselenij v Azerbajdzhane [Excavations of the Eneolithic Settlement in Azerbaijan]', *Arkheologischeskie Otkrytija* 1984, pp.423.

Narimanov, I.G. (1987) *Kul'tura Drevneyshego Zemledel'chesko-Skotovocheskogo Naseleniya Azerbadzhana, Epokho Eneolita VI-IV tys. do n.e.* [The culture of the earliest farmers and herders in Azerbaijan, Chalcolithic period 6th-4 th millenia BC]. Baku, Elm.

Narimanov, I.G. (1992) 'The Earliest Agricultural Settlements in the Territory of Azerbaidzhan', *Soviet Anthropology and Archeology,* 30(4): 9-66.

Niebieridze, L.D. (1986) 'Arkheologischeskie Issledovanija v Tverdzhol'skom Rajone [The Archaeological Investigations in the Tverdzhol Province]', *Polevye Arkheologicheskie Issledovanija* 1983: 8-10.

Nishiaki, Y., Guliyev, F. and Kadowaki ,S. (2015a) 'Chronological Contexts of the Earliest Pottery Neolithic in the South Caucasus: Radiocarbon Dates for Göytepe and Hacı Elamxanlı Tepe, Azerbaijan', *American Journal of Archaeology,* 119: 279-294.

Nishiaki, Y., Guliyev, F., Kadowaki, S., Alakbarov, V., Miki, T., Salimbayov, S., Akashi C. and Arai S. (2015b) 'Investigating Cultural and Socioeconomic Change at the Beginning of the Pottery Neolithic in the Southern Caucasus: The 2013 Excavations at Hacı Elamxanlı Tepe, Azerbaijan', *Bulletin of American Schools of Oriental Research,* 374: 1-28.

Nishiaki, Y. and Le Mière, M. (2005) 'The oldest pottery Neolithic of Upper Mesopotamia: New evidence from Tell Seker al-Aheimar, the Khabur, northeast Syria', *Paléorient,* 31(2): 55-68.

Noy, T. (1985) 'Ten Years of Research in the Area of Gilgal, Lower Jordan Valley', *Israel Museum Journal,* IV: 13-16.

Noy, T., Schuldenrein, J. and Tchernov, E. (1980) 'Gilgal I, a Pre-Pottery Neolithic A Site in the Lower Jordan Valley', *Israel Exploration Journal* , 30: 63.82.

Oates, J. (1969) 'Choga Mami 1967-1968. A Preliminary Report', *Iraq* 31: 115-52.

Oates, J (1987) 'The Choga Mami Transitional' in Huot, J.L. (ed.) *Préhistoire de la Mésopotamie.* Paris: CNRS: 163-80.

Özbaşaran, M. and Duru, G. (2011) 'Akarçay Tepe, A PPNB and PN Settlement in Middle Euphrates – Urfa' in Özdoğan, M., Başgelen, N., Kuniholm, P. (eds) *The Neolithic in Turkey: New Excavations & New Research – The Euphrates Basin (Vol. 2).* Istanbul: Archaeology and Art Publications: 165-202.

Roux, J.C. and Cammas, C. (2010) 'Les techniques constructives en bauge dans l'architecture protohistorique de Latarra (milieu du Ve – milieu du IVe s. av. n. è.)', *Lattara,* 21: 219-288.

Sagona, A. (1993) 'Settlement and Society in Late Prehistoric Trans-Caucasus' in Frangipane M., Hauptmann H., Liverani M., Matthiae P., Mellink M. (eds) *Between the Rivers and Over the Mountains. Archaeologica Anatolica et Mesopotamica Alba Palmieri Dedicata.* Roma: Dipartimiento di Scienze Storiche Archeologiche e Anthropologiche dell'Antichità, Universtà di Roma La Sapienza: 453-74.

Sauvage, M. (1998) *La Brique et sa mise en œuvre en Mésopotamie : des origines à l'époque achéménide.* Paris: ERC.

Sauvage, M. (2001) 'Les briques de grande taille à empreintes de doigts: le Choga Mami Transitional et la culture de Oueili' in Breniquet, C., Kepinski, C. (eds), *Études mésopotamiennes: recueil de textes à Jean-Louis Huot.* Paris: ERC: 417-447.

Sauvage, M. (2009) 'Les débuts de l'architecture de terre au Proche-Orient' in Guillaud, H., Achenza, M. and Corella, M. (eds) *1a Conferenza Mediterranea Sull'architettura in Terra Cruda.* Cagliari: Edicom Edizioni: 189-98.

Smith, J.G. (1952) 'The Matarrah Assemblage', *Journal of the Near Eastern Studies,* 11: 5-56.

Stordeur, D. (2000) *El Kowm 2 : une île dans le désert. La fin du Néolithique précéramique dans la steppe syrienne.* Paris: CNRS.

Stuckenrath, R. (1963) 'University of Pennsylvania Radiocarbon Dates VI', *Radiocarbon,* 5: 82-103.

Suleiman, A. and Nieuwenhuyse, O. (2002) *Tell Boueid II: A Late Neolithic Village on the Middle Khabur (Syria).* Turnhout: Brepols, Subartu, XI.

Tekin, A. (2011) 'Hakemi Use: A Newly Discovered Late Neolithic Site in Southeastern Anatolia' in Özdoğan, M., Başgelen, N. and Kuniholm, P. (eds), *The Neolithic in Turkey: New Excavations & New Research - The Tigris Basin (Vol. 1).* Istanbul: Archaeology and Art Publications: 151-172.
Tobler, A.J. (1950) *Excavations at Tepe Gawra: Joint Expedition of the Baghdad School and the University Museum to Mesopotamia. II. Levels IX-XX.* London: Oxford University Press.
Torosjan, R.M. (1976) *Rannezemledel'cheskoe Poselenie Tekhuta* [A Village of Farmers in Tekhut]. Erevan: Arkeho. Raskopki v Armenii 14.
Verhoeven, M. and Kranendonk, P. (1996) 'The Excavations: Stratigraphy and Architecture' in Akkermans, P. (ed.), *Tell Sabi Abyad: The Late Neolithic Settlement.* Leiden: Nederlands Instituut voor het Nabije Oosten / Istanbul: Nederlands Historisch-Archaeologisch Institut: 25-118.
Voigt, P. (1983) *Hajji Firuz Tepe, Iran: The Neolithic Settlement.* Ann Arbor/London: University Microfilms International.
Wahida, G. (1967) 'The Excavations of the Third Season at Tell es-Sawwan 1966', *Sumer*, 22: 167-76.

## Table 1 Data on cobs and mudbricks

Data are arranged in alphabetical order of the site names. The call number in column one of the table corresponds to the return of the text. Columns three to five specify the cultural attribution, the level, the dating and the building or wall concerned according to the name given by the excavator. Columns eight to ten specify the size of mudbricks (length, width, height), as well as the length/ width ratio and the estimated weight according to these size: the underlined weights correspond to the values greater than 28kg whose manufacture or transport are considered difficult. Columns eleven and twelve specify material used (cob or mudbricks), layout and form (abbreviations below). Finally, the authors' citations as well as the associated bibliographic references are listed in columns thirteen and fourteen. The references preceded by the sign '*' are those of the radiocarbon dates.

### Abbreviations:

C: cob
c: cigar-shaped
E: fingerprints
F: flat
L: lumps
LC: layers of cob
M: moulded mudbricks
m: mudbricks shaped by hands
Ob: oblong
Pl.-cx: plano-convex mudbricks
R: rectangular
und.: undetermined

| N° | Site | Culture/Level | Date | Building | Lenght (in cm) | Width (in cm) | Hight (in cm) | Ratio (L/w) | Weight estimated (in kg) | Technique | Shape or implementation | Citation | Bibliographic references |
|---|---|---|---|---|---|---|---|---|---|---|---|---|---|
| 1 | Akarçay Tepe | Pottery Neolithic phase | 6630-6015 cal. B.C.* | F | 40 | 32 | | 1,25 | | M | | "The kerpiç blocks measure 40 x 32 cm or 32/34 x 22 cm on average." | Özbaşaran et al., 2011:173; *Özbaşaran et al., 2011:167. |
| 2 | | | | | 32-34 | 22 | | | | M | | | |
| 3 | Aknashen-Katunarkh | | | | 50 | 25 | | 2 | | m | F | "The circular buildings demonstrate the presence of clearly evolved darker clay blocks, with dimensions 25 x 50 cm, and 19/20 x 14/15 cm." | Badalyan et al., 2014:165. |
| 4 | | *Aratashen/VII* | Before 6000 B.C.? | | 19 à 20 | 14 | 5 | 1,36 à 1,43 | 2,13 à 2,24 | m | F | | |
| 5 | | *Aratashen/IV* | 5981-5370 cal. B.C.* | | 50 | 30 | 20 | 1,7 | 48 | m | F | | Badalyan et al., 2014:165; *Badalyan et al., 2010:210, tab. 1. |
| 6 | Aratashen | *Aratashen/I* | Before 5800 B.C.? | | 45 | 25 | 8 | 1,8 | 14,40 | m | F | "All are built in flat, rectangular mud bricks of even size (about 45 x 25 x 8 cm). Remains of clay floors are evident in each of them, as well as in the orthogonal structure." | Badalyan et al., 2014:402. |
| 7 | Arpachiyah | *Halaf* | 6400-5400 B.C. | General | | | | | | C | L | "All the tholoi in the Tepe rested on stone foundations, having a superstructure of ***pisé* or beaten clay**." | Mallowan et al., 1935:25. |
| 8 | | *Halaf II/TT7-10* | 5800-5500 B.C. | General | | | | | | C | und. | | |
| 9 | | *Halaf II/TT 4-5* | 5800-5500 B.C. | TT 4-5 | 43 | 41 | 10 | 1,05 | 28.2[illegible] | m | | "In TT 4,5 the following pisé sizes were recovered: 0.43x0.41x0.10 m. [...]. **The size were all irregular.**" (Mallowan *et al.* 1935: 16) | Özbaşaran et al., 2011:173; *Özbaşaran et al., 2011:167. |
| 10 | | *Halaf II/TT 4-5* | 5800-5500 B.C. | TT 4-5 | 37,5 | 21 | 8 | 1,79 | 10,08 | | | | |
| 11 | | | | | 39 | 20,5 | 9 | 1,90 | 11,5[illegible] | | | | |
| 12 | | | | | 46 | 42 | 11 | 1,10 | 34,0[illegible] | | | | |
| 13 | Aruchlo | *Shulaveri-Shomu* | 5900-5750 cal. B.C.* | 9 | | | | | | C | und. | "Le pisé est également connu dans le bassin de la Kura, puisqu'il servit pour quelques édifices circulaires d'Arukhlo 1 (bât. N° 4-9) et de Shulaveri et qu'à Kechili III il est le seul matériau cité." | Chataigner 1995:59; Kushnareva et al,.1970:22; *Lyonnet et al., 2012:85, tab. 19. |
| 14 | | *Shulaveri-Shomu/I* | 5900-5750 cal. B.C.* | 4 | | | | | | C | und. | | |
| 15 | | *Shulaveri-Shomu/Older Neolithic settlement* | 1st third of the 6th Millennium B.C.? | Complex III | 30 à 50 | 20 | 5 | 1,5 à 2,5 | 4,8 à 8 | m | Pl.-cx | "The bricks are 30 to 50 cm long, 20 cm wide and 5 cm in thickness. Normally they are plano-convex; however, the form and the size are variable: they can be smaller and more square." | Hansen et al., 2017:202. |
| 16 | | *Shulaveri-Shomu* | | C030? | 41 | 20 | 8 | 2,05 | 10,50 | m | Pl.-cx | "The individual bricks are usually plano-convex in form, but there is no standard measurement. Nevertheless, in this particular structure some bricks measured 41 cm x 20 cm x 8 cm." (Hansen *et al.* 2013: 390) | Hansen et al., 2006, p.10; Hansen et al., 2013:390. |
| 17 | | | | General | 40 | 15 | 8 | 2,67 | 7,68 | m | | | Dzhavakhishvili, 1973:80; Chubinshvili et al., 1976:58; Munchaev, 1982:106. |
| 18 | | | | General | 40 | 13 | 8 | 3,1 | 6,66 | m | | | |
| 19 | | | | General | 27 | 17 | 10 | 1,6 | 7,34 | m | | | |
| 20 | | | | General | 19 | 17 | 9 | 1,1 | 4,65 | m | | | |
| 21 | | | | General | 18 | 15 | 5 | 1,2 | 2,16 | m | | | |
| 22 | | | | General | 18 | 9 | 7,5 | 2,0 | 1,94 | m | | | |

| N° | Site | Culture/Level | Date | Building | Lenght (in cm) | Width (in cm) | Hight (in cm) | Ratio (L/w) | Weight estimated (in kg) | Technique | Shape or implementation | Citation | Bibliographic references |
|---|---|---|---|---|---|---|---|---|---|---|---|---|---|
| 23 | Aruchlo | *Shulaveri-Shomu* | | General | 32 | 18 | 6 | 1,8 | 5,53 | m | | | Dzhavakhishvili 1973:80; Munchaev, 1982:106. |
| 24 | | | | | 37 | 14 | 11 | 2,6 | 5,50 | M | | | Ioseliani, 2017:281, tab. 1. |
| 25 | | | | | 34 | 17 | 11 | 2,0 | 4,40 | | | | |
| 26 | | | | | 34 | 17 | 10 | 2,0 | 4,60 | | | | |
| 27 | | | | | 34 | 17 | 9 | 2,0 | 4,20 | | | | |
| 28 | | | | | 34 | 16 | 10 | 2,1 | 4,40 | | | | |
| 29 | | | | | 33 | 17 | 12 | 1,9 | 5,30 | | | | |
| 30 | | | | | 32 | 18 | 11 | 1,8 | 4,30 | | | "However due to their standardized size it is possible to assume that at least the majority of the bricks were not produced by hand. Moreover, some of them have **sharp edges and flattened sides suggesting the use of wooden frames**." (Ioseliani 2017: 282) | |
| 31 | | | | | 32 | 18 | 9 | 1,8 | 3,90 | | | | |
| 32 | | | | | 32 | 17 | 12 | 1,9 | 5,30 | | | | |
| 33 | | | | | 32 | 17 | 10 | 1,9 | 4,70 | | | | |
| 34 | | | | | 32 | 16 | 9 | 2,0 | 4,10 | | | | |
| 35 | | | | | 31 | 19 | 11 | 1,6 | 5,60 | | | | |
| 36 | Baghouz | *Late Samarra* | | General | 59 | 24 à 29 | 8 à 9 | 2,03 à 2,46 | 18,12 à 24,64 | M | | «**Briques moulées**, leurs dimensions correspondraient respectivement à deux coudées sur une (59 x 26,5 cm) et à une coudée et demie sur une (45 x 31 cm)" (Sauvage 2001: 436) | Du Mesnil Du Buisson, 1948:15; Sauvage, 2001, p.436. |
| 37 | | | | | 36 | 30 | 9 | 1,20 | 15,55 | | | | |
| 38 | | | | | 45 | 31 | 10 | 1,45 | 22,32 | | | | |
| 39 | Boueid II | *Proto-Hassuna* or *Pre-Halaf* | 6500-6300 B.C. | General | | | | | | C | und. | "The buildings were all built of **pisé** (...)." | Suleiman et al. 2002:6. |
| 40 | | | | | 30 | 25 | 8 | 1,20 | 9,60 | m | R | "The buildings were all built of pisé, although mud bricks may have been used occasionally, in the western and the southern wall of building I. These bricks **do not appear to have a standardized size** (varying around 30x25x8 cm)." | |
| 41 | Bouqras | Late PPNB/II | 7036-6605 cal. B.C.* | General | 54 | 27 | 7 | 2,00 | 16,33 | M | | «Les murs sont désormais construits en briques quadrangulaires **faites au moule**, de 7 cm d'épaisseur et de 25 à 35 cm de côté." | Contenson, 1985:338; *Hours *et al.*, 1994, pp. 388-389. |
| 42 | | | | | 30 | | 7 | | | | | | |
| 43 | | Late PPNB/III | 6996-6472 cal. B.C.* | | 30 | 30 | | | 1 | | | | |
| 44 | Chagar Bazar | *Proto-Halaf/* phase 10 | 6000-5600 cal. B.C.* | *Circular building* | 50 | 20 | 8 | 2,50 | 12,80 | M | | "In general, their walls were erected in **moulded** or hand-shaped mud bricks [...]. This building was constructed with mud bricks measuring 25 cm x 50 cm x 8 cm." | Cruells et al., 2013:471; *Cruells et al., 2013:393, tab. 42.1. |

| N° | Site | Culture/Level | Date | Building | Lenght (in cm) | Width (in cm) | Hight (in cm) | Ratio (L/w) | Weight estimated (in kg) | Technique | Shape or implementation | Citation | Bibliographic references |
|---|---|---|---|---|---|---|---|---|---|---|---|---|---|
| 45 | Choga Mami | *Late Samarra/ III* | | | 60 à 90 | 12 à 18 | 15 | 3,33 à 7,5 | 17,28 à 38,88 | m | c | "The mud-brick employed in this building and elsewhere under on the site is long, cigar-shaped, and sun-dried, 60-90 cm. in lenght and from 12-18 cm. in diameter." | Oates, 1969:116. |
| 46 | | *Halaf/II* | | | 55 | | | | | | c; E | «Like the surrounding wall, these were constructed of **cigar-shaped libn** (55 cm. long)." | Oates, 1969:117. |
| 47 | Choga Mish | *Archaic Susiana 3* | | | 80 à 95 | 12 à 20 | 10 à 12 | 4 à 7,92 | 15,36 à 36,48 | m | E | | Delougaz et al., 1972:93. |
| 48 | | | | | 50 à 60 | 11 à 12 | 11 à 12 | 4,17 à 5,5 | 9,68 à 13,82 | | | | |
| 49 | Choga Sefid | *Choga Mami Transitional (CMT)* | | | 80 | 10 | 10 | 8 | 12,80 | m | E | | Hole, 1977:78. |
| 50 | | | | | 75 | 15 | 15 | 5 | 27,00 | | | | |
| 51 | | | | | 160 à 260 | 15 | 15 | 10,67 à 17,33 | 57,60 à 93,60 | | | | |
| 52 | Dalma Tepe | *Dalma Tepe/5-4* | 5068-4698 cal. B.C.* | General | | | | | | C | LC | "The walls at Dalma were constructed of *chineh* **(layers of packed mud)**, separated in some cases by thin mortar-like clay layers, and were preserved to heights ranging from 15-60 cm." | Hamlin, 1975:113; *Hamlin, 1975:119, tab. 2. |
| 53 | Fıstıklı Höyük | *Halaf I/II* | 6000-5300 B.C. | General | | | | | | C | und. | "The only material used for the walls was pisé [...]. These two kinds of pisé could be observed: a crumbly reddish-brown pisé with chaff inclusions, and a more homogeneous, silty, tannish-colored material with little chaff." | Bernbeck et al., 2003:26. |
| 54 | Gadachrili Gora | *Shulaver-Shomu/II* | 5900-5300 cal. B.C.* | Wall 217 | | | | | | C | L | "These **blocks** are composed of clay mixed with organic inclusions (charcoal, bones, etc). **The clay used was sometimes sourced directly from anthropic levels or former soils**, which make it difficult to distinguish the walls from the surrounding destruction or fill layers (see for example Wall 217) (Fig. 23)." | Hamon et al., 2016:164; *Hamon et al. 2016:164, fig. 4. |
| 55 | | *Shulaveri-Shomu/I* | | 2002 | 20 à 26 | 12 à 14 | 8 | 1,43 à 1,67 | 3,07 à 3,58 | m | Pl-cx | | Hamon et al., 2016, pp. 160, 162-163; Personal data; *Hamon et al., 2016:1964, fig. 4. |
| 56 | | | | 2003 | 19 | 19 | 5 | 1 | 2,89 | | F | | |
| 57 | | | | | 40 | 19 | 9 | 2,11 | 10,94 | | Pl-cx | | |

| N° | Site | Culture/Level | Date | Building | Lenght (in cm) | Width (in cm) | Hight (in cm) | Ratio (L/w) | Weight estimated (in kg) | Technique | Shape or implementation | Citation | Bibliographic references |
|---|---|---|---|---|---|---|---|---|---|---|---|---|---|
| 58 | Gadachrili Gora | Shulaveri-Shomu/I | 5900-5300 cal. B.C*. | 2004 | 50 | 25 | | 2 | | m | F | "The building was constructed using elongated, flat, rectangular mud brick laid in a regular stretcher bond." (Hamon *et al.* 2016: 159) | Hamon et al., 2016:159; Personal data; *Hamon et al., 2016, p.164, fig. 4. |
| 59 | | | | | 38 à 40 | 15 à 20 | 8 à 10 | 2,53 à 2,67 | 7,3 à 12,8 | | | | Personal data; *Hamon et al., 2016:164, fig. 4. |
| 60 | | *Shulaveri-Shomu* | | Wall 234 | 23 | | 7 | | | | | | |
| 61 | Gargalar Tepesi | *Shulaveri-Shomu/3* | | 4 | 36 | 16 | 8 | 2 | 7,37 | m | | "Utility structure no. 4, 1.8 m in diameter, was built from brick of 36 x 16 x 8 cm (brick of dimensions 50 x 20 x 8-10 cm is also encountered in the third horizon)" | Narimanov, 1992:21. |
| 62 | | | | 9 | 26 à 40 | 15 | 7 à 8 | 1,7 à 2,7 | 4,37 à 7,68 | | | | |
| 63 | | | | General | 50 | 20 | 8 à 10 | 3 | 12,8 à 16 | | | | |
| 64 | | *Shulaveri-Shomu/2* | | | 44 | 16 à 18 | 7 à 9 | 2,44 à 2,75 | 7,89 à 11,41 | | | | |
| 65 | | | | | 40 | 16 | 8 à 11 | 3 | 8,19 à 11,26 | | | | |
| 66 | | | | | 34 | 14 | 8 | 2 | 6,09 | | | | |
| 67 | | *Shulaveri-Shomu/1* | | | 46 à 48 | 16 à 18 | 7 à 9 | 2,56 à 3 | 8,24 à 12,44 | | | "The bricks used to build the structures in the first horizon […], and they were made of clay with chopped chaff; only the bricks of structure no. 20 were made of pure clay." | Narimanov, 1992:20. |
| 68 | | | | | 40 | 16 | 8 | 3 | 8,19 | | | | |
| 69 | Gawra | *Halaf/XX* | | | 13 | 9 | 5 | 1,44 | 0,94 | M | | "The bricks forming them were the smallest yet discovered, measuring **as a rule** about 13 x 9 x 5 cm." | Tobler, 1950:48. |
| 70 | Göy Tepe | *Shulaveri-Shomu/1 à 14* | 5645-5460 cal. B.C.* | General | 30 à 50 | 15 à 18 | 10 | 1,67 à 3,33 | 7,2 à 14,4 | m | Pl-cx | "The walls are made of plano-convex mud bricks with straw temper, either in a yellow or grey color." | Guliyev et al., 2014:5; *Nishiaki et al., 2015:286, tab. 1. |
| 71 | | | | | 60 | 20 | 8 | 3 | 15,36 | | | "Hündür divarlı tikilerde kerpiclerin ölçüsü 60x20x8x8 sm, hörgüsü 12-15 cergeden ibaretdir." | Guliyev et al., 2009:47; *Nishiaki et al., 2015:286, tab. 1. |
| 72 | | | | | 40 à 60 | 20 | 8 à 10 | 2 à 3 | 10,24 à 19,2 | | | "The building materials were plano-convex-shaped mud bricks with straw temper, measuring approximately 40-60 x 20 x 8-10 cm. The bricks were made with yellow or brown clay." | Guliyev et al., 2012:74; *Nishiaki et al., 2015:286, tab. 1. |
| 73 | Gritille | PPNB/Late PPNB | 7500-6380 cal. B.C.* | General | 80 | 35 | | | 2,29 | M? | | | Aurenche et al., 1988:5; *Hours et al., 1994:396. |
| 74 | | | | | 45 | 35 | | | 1,29 | | | | |
| 75 | | | | | 35 | 35 | | | 1 | | | | |

| N° | Site | Culture/Level | Date | Building | Lenght (in cm) | Width (in cm) | Hight (in cm) | Ratio (L/w) | Weight estimated (in kg) | Technique | Shape or implementation | Citation | Bibliographic references |
|---|---|---|---|---|---|---|---|---|---|---|---|---|---|
| 76 | Haji Firuz | *Hajji Firuz*/J | 6225-5528 cal. B.C.* (phase *Hajji Firuz*) | Str. XV2 | | | | | | C | LC | "Structures I, and V, and perhaps the buildings dating to Phases L through G (see below), were constructed of **packed mud**. In this method of construction the clay is mixed with water and a tempering agent, and then **packed down in a layer which runs the entire length of the wall. After the layer dries *in situ*, the process is repeated until the desired height is reached**." | Voigt, 1983:33; *Lawn, 1974:222. |
| 77 | | *Hajji Firuz*/G | | Str. XII1 | | | | | | | | | |
| 78 | | *Hajji Firuz*/F | | Str XI1 | | | | | | | | | |
| 79 | | *Hajji Firuz*/C | | Str. V | | | | | | | | | |
| 80 | | *Hajji Firuz*/ A2-A3 | | Str. I1 | | | | | | | | | |
| 81 | | *Hajji Firuz*/F (then G) | | Str XI1 | 12 à 58 | | 6 | | | M | | "It is therefore unlikely that the bricks were **made in a frame or mold**. Clay mixed with water and vegetable matter (probably grass or straw) was presumably shaped into relatively thin slabs by hand, sun dried, and then set into mortar to form the walls." | Voigt, 1983:33; *Lawn, 1974:222. |
| 82 | | *Hajji Firuz*/C | | Str. VI | 39 | 24 | 8 | 1,63 | 11,98 | | | | Voigt, 1983:48; *Lawn, 1974:222. |
| 83 | | | | | 28 | 30 à 33 | 5 à 8 | 0,85 à 0,93 | 6,72 à 11,83 | | | | Voigt 1983:47; *Lawn, 1974:222. |
| 84 | | | | | 34 | 30 à 33 | 5 à 8 | 1,03 à 1,13 | 8,16 à 14,36 | | | | |
| 85 | | | | | 44 | 30 à 33 | 5 à 8 | 1,33 à 1,47 | 10,56 à 18,59 | | | | |
| 86 | | | | Str. I1 | 44 | 40 | | 1,10 | | | | | *Lawn, 1974:222 |
| 87 | | *Hajji Firuz*/B-C | | Str. II2 | 36 | 24 à 26 | 6 | 1,39 à 1,5 | 8,29 à 8,99 | | | | Voigt, 1983:42; *Lawn, 1974:222. |
| 88 | | *Hajji Firuz*/ Phase A3 | | Str. III | 37 à 40 | 24 à 26 | 6 | 1,42 à 1,67 | 8,52 à 9,98 | | | | Voigt, 1983:43; *Lawn, 1974:222. |
| 89 | Hakemi Use | *Hassuna* or *Samarra*/III | 6000-5900 B.C. | General | | | | | | C | und. | "The use of the *pisé* technique and the loosely bound texture of the walls is another common feature in the architecture of the site [...]. The majority of the walls in the buildings were produced using *pisé*, while some parts of the walls were constructed with adobe parts [...]. This oddity is observed at the upper level and only in three buildings." | Tekin, 2011, pp. 152-153. |

| N° | Site | Culture/Level | Date | Building | Lenght (in cm) | Width (in cm) | Hight (in cm) | Ratio (L/w) | Weight estimated (in kg) | Technique | Shape or implementation | Citation | Bibliographic references |
|---|---|---|---|---|---|---|---|---|---|---|---|---|---|
| 90 | Hakemi Use | *Hassuna-Samarra*/III | 6000-5900 B.C. | General | 45 | 45 | 10 | 1 | 32,40 | m | | "The dimensions of the mudbricks are not standardized; the largest of them are 45 x 45 x 15 cm; the smallest ones 30 x 15 x 10 cm. It should be noted that these mudbricks are produced using a simple technique and have a thin surface tissue possibly due to only a short period's being allowed for drying process." | Tekin, 2011:152. |
| 91 | | | | | 30 | 15 | 10 | 2 | 7,20 | | | | |
| 92 | Hassan | *Halaf II* | 5800-5500 B.C. | *Tholos* | | | | | | C | und. | "It is a small *tholos* less than 3m. In diameter, of which only the ***tauf*** base remains." | Jasim, 1985:164. |
| 93 | | *Hassuna*/I-II | 6600-5800 B.C. | General | | | | | | | L | **"lumps of mud of various sizes"** | Safar et al., 1945:273. |
| 94 | Imiris Gora | *Shulaveri-Shomu*/IV | 5730-4950 cal. B.C.*? | 35 | 38 | 16 | 11 | 2,38 | 10,70 | m | Pl-cx | | Dzhaparidze et al., 1971:28; *Dzhaparidze et al., 1975:127. |
| 95 | | | | | 41 | 16 | 10 | 2,56 | 10,50 | | | | |
| 96 | | *Shulaveri-Shomu* | | General | 50 | 20 | 10 | 2,50 | 16,00 | | | | Dzhavakhishvili, 1973:48. |
| 97 | | | | | 40 | 15 | 10 | 2,67 | 9,60 | | | | |
| 98 | | | | | 38 | 15 | 10 | 2,53 | 9,12 | | | | |
| 99 | | | | | 35 | 16 | 10 à 11 | 2,19 | 8,96 à 9,86 | | | | |
| 100 | | | | | 32 | 20 | 7 | 1,60 | 7,17 | | | | |
| 101 | Kamiltepe | *Kamiltepe*/Phase Kamiltepe I | 5610-5380 cal. B.C.*? | | 15 | 16 | | 1 | | m | | "The construction of the platform was made with **hand-shaped** bricks of irregular shape [...]. The bricks were oblong, with an approximately square section; after the slabs has been dug out, they had apparently been smoothed along the long sides by hand, **and then the front ends were compressed to a flat surface by hand**." (Helwing *et. al.* 2017: 17) | Aliyev et al., 2009:29; Helwing et al., 2017:17; *Aliyev et al., 2009:38, fig. 21. |
| 102 | | | | | 19 | 20 | | 1 | | | | | |
| 103 | | | | | 34 | 18 | 15 | 2 | 14,69 | | | | Narimanov, 1992:35; *Aliyev et al., 2009:38, fig. 21. |
| 104 | Khramis Didi-Gora | *Shulaveri-Shomu*/III | Before 5300 B.C.? | 27 | 42 | 20 | 7 à 7,5 | 2,10 | 9,41 à 10,08 | m | Pl.-cx? | | Personal data. |
| 105 | Khramis Didi-Gora | *Shulaveri-Shomu*/I | Before 5300 B.C.? | 34 | 42 | 16 | 7 | 2,63 | 7,53 | m | Pl.-cx? | | Personal data. |
| 106 | | | | 34 | 48 à 49 | 24 | 7 | 2 à 2,04 | 12,9 à 13,17 | | | | |
| 107 | | | | 29 | 42 à 43 | 20 à 22 | 7,5 | 1,9 à 2,15 | 10,08 à 11,35 | | | | |
| 108 | | *Shulaveri-Shomu* | | General | 48 | 24 | 7 | 2 | 12,90 | | Pl.-cx | | Kiguradze, 1986:70. |
| 109 | | | | | 45 | 20 | 7 | 2,25 | 10,08 | | | | |
| 110 | | | | | 42 | 20 | 7 | 2,10 | 9,41 | m | | | |
| 111 | | | | | 36 | 18 | 7,5 | 2 | 7,78 | | | | |
| 112 | Kültepe | *Kültepe* | 6000-5300 B.C.? | General | | | | | | C | und. | | Abibullaev, 1963, pp. 157-158. |

| N° | Site | Culture/Level | Date | Building | Lenght (in cm) | Width (in cm) | Hight (in cm) | Ratio (L/w) | Weight estimated (in kg) | Technique | Shape or implementation | Citation | Bibliographic references |
|---|---|---|---|---|---|---|---|---|---|---|---|---|---|
| 113 | Masis Blur | *Aratashen*/I | 5925-5630 cal. B.C.* | S003, S004, S005 | | | | | | C | L | "The second technique describes preparation of **mud "clods" or "pads"** (*motte – fr.*) from the mixture of clay tempered with cut straw (*torchis – fr.*)." | Hayrapetyan, et al., 2014:180; *Martyrosyan-Olshansky et al., 2013:145, tab. 1. |
| 114 | | | | S011 | | | | | | | LC | "One of the structures (S011) differs from the aforementioned ones and represents **a straight row of alternating dark and light clay rectangles (pisé and clods)**." | |
| 115 | Matarrah | *Hassuna*/I-II | 6300 - 5800 B.C. | General | | | | | | C | und. | "The houses of this level were of four to six rooms, constructed of **packed mud or *tauf*.**" | Braidwood et al., 1952:6. |
| 116 | Mentesh Tepe | *Shulaveri-Shomu*/I | 5800-5600 cal. B.C.* | 285=689 | 44 à 45 | 15 à 16 | 9 à 10 | 2,75 à 3 | 9,5 à 11,52 | M | Pl.-cx | | Personal data; *Lyonnet et al., 2016:180, tab. 2. |
| 117 | | | | 293 | 43 | 15 | 10 | 2,87 | 10,32 | | | | |
| 118 | | | | 346 | 28 | 22 | 8 | 1,27 | 7,83 | m | | | |
| 119 | | | | 284 | 43 | 14 | 9 | 3,07 | 8,67 | | Pl.-cx | | |
| 120 | | | | 516 | 43 | 14 | 9 | 3,07 | 8,67 | | | | |
| 121 | | | | 718=337 | 40 | 15 | 10 | 2,67 | 9,3 | M? | | | |
| 122 | | | | 1025 | 43 | 12 | 10 | 3,58 | 8,26 | m | | | |
| 123 | | | | 1031 | 39 | 13 | 10 | 3 | 8,11 | M? | | | |
| 124 | Sabi Abyad | *Pre-Halaf*/8 | 6125-6075 cal. B.C.* | 8.1 | | | | | | C | L | "made of reddish, crumbly earth (representing either ***pisé* or difficult-to-recognize layers of slabs**) mixed with small pieces of limestones" | Akkermans et al., 2014, pp. 36-37; *Akkermans et al., 2014:31, tab. 2.2. |
| 125 | | | | 8.2 | | | | | | | und. | "The 35 cm-wide walls were made of large **clay slabs up to one meter long** and covered with brownish mud plaster up to 3 cm thick." | |
| 126 | Sabi Abyad | *Pre-Halaf*/7A | 6015-5995 cal. B.C.* | 7.5 | | | | | | C | L | "The eastern wall bounding the rectangular annex was constructed in two stages: its lower part consisted of compact brown clay, including irregular, reddish-brown wall pieces, to a height of about 50 cm, whereas its upper part was made of greyish clay layers, preserved to a height of about 60 cm. The use of different building materials may have been simply due to the simultaneous exploitation of various sources of clays for construction purposes." | Akkermans et al., 2014:57; *Akkermans et al., 2014:31, tab. 2.2. |
| 127 | | *Pre-Halaf*/7B | 6020-6005 cal. B.C.* | 7.3 | | | | | | C | LC | "Its wall about **40 cm thick was made of compact brown clay"** | Akkermans et al., 2014:52; *Akkermans et al., 2014:31, tab. 2.2. |
| 128 | | | | 7.1 | | | | | | | und. | "They were made either of **large slabs or of compact clay** often including slab fragments **(re-used pieces** of earlier walls), or of combination of these." | Akkermans et al., 2014, p.45; * Akkermans et al., 2014:31, tab. 2.2. |

| N° | Site | Culture/Level | Date | Building | Lenght (in cm) | Width (in cm) | Hight (in cm) | Ratio (L/w) | Weight estimated (in kg) | Technique | Shape or layout | Citation | Bibliographic references |
|---|---|---|---|---|---|---|---|---|---|---|---|---|---|
| 129 | Sabi Abyad | *Pre-Halaf*/7-6 | 6020-5995 cal. B.C.* | 7.5 | 45 à 50 | 35 | 6 à 8 | 1,29 à 1,43 | 15,12 à 22,4 | m? | | "Most walls seemed to consist of **regular** layers of grey-brown mud bricks, measuring about 45/50 x 35 x 6-8 cm, alternating with layers of a dark-brown mortar 1-3 cm thick." | Akkermans et al., 2014:57; * Akkermans et al., 2014:31, tab. 2.2. |
| 130 | | | | 7.11 | 40 à 44 | 40 à 45 | 4 à 7 | 0,88 à 1,1 | 10,24 à 22,18 | | | "They were mainly made of brown mud bricks, measuring 40-44 by 40-45 by 4-7 cm" | Akkermans et al., 2014:61; * Akkermans et al., 2014:31, tab. 2.2. |
| 131 | | | | 6.13 | 40 | 40 | 4 à 6 | 1 | 10,24 à 15,36 | | | "Extensive remains of this wall occurred due north outside the tholos, in the form of collapsed stretches of roughly square, reddish-brown to grey-brown bricks (each about 40 by 40 by 4-6 cm) set in a grey mud mortar." | Akkermans et al., 2014:74; * Akkermans et al., 2014:31, tab. 2.2. |
| 132 | Sabi Abyad | *Early Halaf*/6 | 6010-5995 cal. B.C.* | 6.1, 6.2, 6.4, 6.5, 6.6, 6.7, 6.8, 6.9, 6.10, 6.11 | | | | | | C | LC | "The generally 40 cm wide walls of the level 6 buildings were simply founded on earth and were all built of pisé, laid down in **layers of various coulours and, most likely, various consistency**. The order was always the same : a grey, 2 or 3 cm thick loam band followed by an orange-brown, ca. 2-4 cm thick deposit, in its turn covered by a buff layer, about 6-8 cm thick, etc. Apparently, various sources of clay were in use, which each must have had different qualities." | Verhoeven et al., 1996:44; * Akkermans et al., 2014:31, tab. 2.2. |
| 133 | | | | | | | | | | | | | |
| 134 | | *Early Halaf*/5 | 5995-5925 cal. B.C.* | 5.1 | 40 | 30 à 35 | 8 | 1,14 à 1,33 | 15,36 à 17,92 | m? | | "The walls were each ca. 25-50 cm thick and built either of compact grey pisé or **grey mud bricks** measuring ca. 40 x 35/30 x 8 cm." | Verhoeven et al., 1996:66; * Akkermans et al., 2014:31, tab. 2.2. |
| 135 | | | | 5.2 | 30 | 30 | 12 | 1 | 17,28 | | | | * Akkermans et al., 2014:31, tab. 2.2. |
| 136 | | *Halaf*/3B | 5940-5905 cal. B.C.* | Tholos I | | | | | | C | LC | "The walls were built of pisé, which was laid down in **two alternating layers of orange-brown bands** ca. 6-8 cm thick and grey bands ca. 2 cm thick." | Verhoeven et al., 1996:94; * Akkermans et al., 2014:31, tab. 2.2. |
| 137 | | *Early Halaf*/3B | | 1 | 35 à 40 | 35 | 10 | 1 à 1,14 | 19,6 à 22,4 | m? | | "Commonly, large and more or less square mud bricks measuring ca. 35/40 x 35 x 10 cm were used as well." | Verhoeven et al, 1996:92; * Akkermans et al., 2014:31, tab. 2.2. |
| 138 | | | | Tholos S | 35 | 30 | 8 | 1,17 | 13,44 | | | "Tholos S had an interior diameter of about 2.85 m and stood to a height of ca. 40 cm. It was built of mud bricks each measuring ca. 35 x 30 x 8 cm." | Verhoeven et al., 1996:96; * Akkermans et al., 2014:31, tab. 2.2. |
| 139 | | | | Tholos I | 30 | 25 | 8 | 1,20 | 9,60 | | | "The tholos gave evidence of a mud-brick wall, built of bricks measuring ca. 30 x 25 x 8/10 cm." | |
| 140 | | | | Tholos N/ AE | 35 | 30 | 8 | 1,17 | 13,44 | | | "This second tholos was built of mud bricks each measuring about 35 x 30 x 8 cm, and joined by a grey mortar ca. 2 cm thick." | Verhoeven et al., 1996:97; * Akkermans et al., 2014, p.31, tab. 2.2. |

| N° | Site | Culture/Level | Date | Building | Lenght (in cm) | Width (in cm) | Hight (in cm) | Ratio (L/w) | Weight estimated (in kg) | Technique | Shape or layout | Citation | Bibliographic references |
|---|---|---|---|---|---|---|---|---|---|---|---|---|---|
| 141 | Salat Cami Yani | Level 2 | 6600-6100 cal. B.C.* | General | | | | | | C | und. | "The *pisé* building were built without stone foundations." | Miyake, 2011:131; Miyake, 2011:149, fig. 30. |
| 142 | | | | 1 | 60 | 25,5 | 7 | 2,35 | 17,14 | | | | Youkana, 1997:15; *El-Wailly et al., 1965:19. |
| 143 | | | | | 90 | 30 | 8 | 3 | 34,56 | | | | |
| 144 | | | | | 105 | 30 | 8 | 3,50 | 40,32 | | | | |
| 145 | | PN(?)/I | | | 100 | 30 | 8 | 3,33 | 38,40 | | | | |
| 146 | | | | 3 | 90 | 30 | 8 | 3 | 34,56 | | | | Youkana, 1997, p.43; * El-Wailly et al., 1965:19. |
| 147 | | | | | 94 | 30 | 8 | 3,13 | 36,10 | | | | |
| 148 | | | 6420-5525 cal. B.C.* | | 74 | 30 | 8 | 2,47 | 28,42 | M | | | |
| 149 | | | | | 70 | 30 | 8 | 2,33 | 26,88 | | | | |
| 150 | Sawwan | PN(?)/I-II | | Trench 4 | 48 | 24 | 8 | 2 | 14,75 | | | "Des lambeaux d'une installation apparurent à la surface: les restes d'un gros mur constitué d'au moins deux rangs de briques crues (48 x 24 x 8 cm) disposées en alternance boutisses/panneresses, orienté nord-ouest/sud-est et doté d'un parement de briques de chant et peut-être d'un contrefort de briques disposées dans le sens de la longueur, à l'ouest." | Breniquet, 1992:17. |
| 151 | | PN(?)/II<br>5975-5704 cal. B.C.*? | | | 80 | 30 | 12 | 2,67 | 46,08 | | Ob | "As in level I the walls were constructed of large rectangular sun-dried, mud-brick, chaff-tempered and moulded [...]. Some of the *libn* are extremely long, and in one room, or even one wall, there is a considerable variation in the lenght of the *libn*. The average size is approximately 80x30x8 cm [...]. The large elongated mud-brick used at Tell es-Sawwan would have presented considerable building problems: if moulded and dried on the ground they would have been very heavy and difficult to lift on to the walls yet if they had been moulded and allowed to dry in position, there would have been no need for mortar." | |
| 152 | | | | | 60 à 100 | 30 à 34 | 6 à 8 | 1,76 à 3,33 | 17,28 à 48,96 | m | | | :Wahida, 1967:172; *Burleigh et al., 1982247. |

| N° | Site | Culture/Level | Date | Building | Lenght (in cm) | Width (in cm) | Hight (in cm) | Ratio (L/w) | Weight estimated (in kg) | Technique | Shape or layout | Citation | Bibliographic references |
|---|---|---|---|---|---|---|---|---|---|---|---|---|---|
| 151 | | | | | 80 | 30 | 12 | 2,67 | 46,08 | | Ob | "As in level I the walls were constructed of large rectangular sun-dried, mud-brick, chaff-tempered and moulded […]. Some of the *libn* are extremely long, and in one room, or even one wall, there is a considerable variation in the lenght of the *libn*. The average size is approximately 80x30x8 cm […]. The large elongated mud-brick used at Tell es-Sawwan would have presented considerable building problems: if moulded and dried on the ground they would have been very heavy and difficult to lift on to the walls yet if they had been moulded and allowed to dry in position, there would have been no need for mortar." | :Wahida, 1967:172; *Burleigh et al., 1982247. |
| 152 | | PN(?)/II | 5975-5704 cal. B.C.*? | | 60 à 100 | 30 à 34 | 6 à 8 | 1,76 à 3,33 | 17,28 à 48,96 | m | | | |
| 153 | | | | | 94 | 30 | 9 | 3,13 | 40,61 | | | | |
| 154 | | | | | 98 | 34 | 8 | 2,88 | 42,65 | | | | |
| 155 | | | | | 64 | 34 | 9 | 1,88 | 31,33 | | | | |
| 156 | | | | | 63 | 28 | 7 | 2,25 | 19,76 | | | | |
| 157 | | | | 2 | 73 | 28 | 8 | 2,61 | 26,16 | | | | Youkana, 1997: 43; *Burleigh et al., 1982:247. |
| 158 | | | | | 65 | 28 | 8 | 2,32 | 23,30 | | | | |
| 159 | | | | | 58 | 30 | 8 | 1,93 | 22,27 | | | | |
| 160 | Sawwan | | | | 73 | 30 | 7 | 2,43 | 24,53 | | | | |
| 161 | | | | | 94 | 30 | 8 | 3,13 | 36,10 | | | | |
| 162 | | | | | 60 | 30 | 9 | 2 | 25,92 | | | | |
| 163 | | | | | | | | | | | | | |
| 163 | | *Samarra* | | Trench 2 | 53 | 25 | | 2,12 | | M | | «Ces briques étaient de dimensions communes (53 x 25 x ? cm) et leur appareillage était encore discernable. Sur l'assise conservée, elles étaient placées en boutisses. Au nord-est, une rangée de briques de chant paraissait plaquée contre le mur (Fig. I)." | Breniquet, 1992:19. |
| 164 | | | | | 69 | | 8 | | | | | «Légèrement en retrait de cette ouverture vers le sud-est, une structure (n°9) de nature indéterminée formée de six rangées de briques (69 x ? X 8 cm) disposées sur la tranche fut dégagée." | Breniquet, 1992:21. |
| 165 | | *Samarra*/IV | | Général | 50 à 70 | 21 à 30 | 6 à 8 | 1,67 à 3,33 | 10,08 à 26,88 | | | "They were all constructed of rather large oblong mudbricks, **made in moulds** and measuring 50-70 by 21-30 by 6-8 cm. (fig. 56); the thickness of a wall is the normal width of a brick (21-30 cm)." | El-Wailly, 1965:21. |
| 166 | | | | 12 | 50 | 25 | 8 | 2 | 16 | | | «Si le mur 38 semble avoir une largeur constante de deux largeurs de briques (environ 50 x 25 x 8 cm), il n'en va pas de même pour le mur 37 […]." | Breniquet, 1992:9. |

| N° | Site | Culture/Level | Date | Building | Lenght (in cm) | Width (in cm) | Hight (in cm) | Ratio (L/w) | Weight estimated (in kg) | Technique | Shape or layout | Citation | Bibliographic references |
|---|---|---|---|---|---|---|---|---|---|---|---|---|---|
| 167 | Shomu Tepe | *Shulaveri-Shomu* | 6520-6200 cal. B.C.*? | General | 50 à 55 | 22 à 25 | 8 | 2 à 2,5 | 14,08 à 17,6 | m | Pl.-cx | | Chataigner, 1995:72; *Kiguradze, 1986:112, tab. 5. |
| 168 | | | | | 36 | 16 | 10 | 2,25 | 9,22 | | | | Narimanov, 1987:86; *Kiguradze, 1986:112, tab. 5. |
| 169 | | | | | 34 à 36 | 15 à 16 | 8 | 2,13 à 2,4 | 6,53 à 7,37 | | | | |
| 170 | | | | | 34 | 13,5 | 9 | 2,52 | 6,61 | | | | |
| 171 | | | | | 32 à 35 | 13 | 8 | 2,46 à 2,69 | 4,57 à 6,74 | | | | |
| 172 | | | | | 32 à 34 | 16 | 8 à 9 | 2 à 2,3 | 6,56 à 7,84 | | | | |
| 173 | | | | | 32 à 33 | 15 à 16 | 8 à 9 | 2,13 à 2,2 | 6,14 à 7,6 | | | | |
| 174 | | | | | 32 | 16 | 8 à 9 | 2 | 6,55 à 7,37 | | | | |
| 175 | | | | | 32 | 15 | 8 | 2,1 | 6,14 | | | | |
| 176 | Shulaveris Gora | *Shulaveri-Shomu* | 5745-5365 cal. B.C.* | General | 50 | 20 | 8 | 2,5 | 12,80 | m | Pl.-cx | | Dzhavakhishvili, 1973:19; *Kiguradze, 1976:168. |
| 177 | | | | | 44 | 25 | 10 | 1,8 | 17,60 | | | | |
| 178 | | | | | 43 | 25 | 7 | 1,7 | 12,04 | | | | |
| 179 | | | | | 41 | 15 | 7 | 2,7 | 6,89 | | | | |
| 180 | | | | | 35 à 37 | 15 | 7 | 2,33 à 2,47 | 5,88 à 6,22 | | | | |
| 181 | | | | | 30 | 20 | 8 | 1,5 | 7,68 | | | | |
| 182 | | | | | 30 | 15 | 8 | 2 | 5,76 | | | | |
| 183 | | | | | 28 | 20 | 8 | 1,40 | 7,17 | | | | |
| 184 | | | | | 25 | 20 | 8 | 1,25 | 6,40 | | | | |
| 185 | | | | | 25 | 15 | 7 | 1,67 | 4,20 | | | | |
| 186 | | | | | | | | | | C | und. | "Buildings at Shulaveris were constructed mainly of plano-convex, sun dried mud brick and some wattle-and-daub, and erected without stone foundation." (Sagona 1993: 456) | Chataigner, 1995:59; Dzhaparidze et al., 1975:203; Sagona, 1993:456; *Kiguradze, 1976:168. |

| N° | Site | Culture/Level | Date | Building | Lenght (in cm) | Width (in cm) | Hight (in cm) | Ratio (L/w) | Weight estimated (in kg) | Technique | Shape or layout | Citation | Bibliographic references |
|---|---|---|---|---|---|---|---|---|---|---|---|---|---|
| 187 | Sungur A | *Late Samarra* | *Beginning of the 6th Millennium B.C.?* | 1 to 6 | 70 | 16 | 8 | 4,38 | 14,34 | m | E | | Matsumoto, 1984:37; Matsumoto, 1987:189. |
| 188 | | | | | 72 | 18 | 8 | 4 | 16,59 | | | | |
| 189 | Sungur B | *HUT/II* | 5400-5300 B.C.? | B1 | | | | | | C | und. | "B1 measured 23 x 10 m. and 70 cm. wide, and is built of *tauf.*" (Jasim 1985: 155); «The wall made of **"tauf"** is 70 cm wide, and is poor in preservation [...]." (Fuji 1981: 183) | Jasim, 1985:155; Fuji, 1981:183. |
| 190 | | | | B2 | | | | | | | | "The other building, B2, is symmetrically constructed with cross-shaped [...]. It is built of ***tauf*** about 80 cm. thick." (Jasim 1985: 155); "The wall made of "tauf" is about 80 cm thick." (Fuji 1981: 183-184) | |
| 191 | | *HUT/III* | | B1/B2 | | | | | | | | "The **"tauf"wall** is 1.0-0.7 m thick..." | Fuji, 1981:184 |
| 192 | Sungur B, C | *Halaf* | | General | | | | | | | | **""Tauf" is used for the building walls.** Despite the fact that mud bricks were adopted in the Samarran buildings discovered at Songor A, this mode is not taken over into the Ubaid and Halaf Periods." | Fuji, 1981:190. |
| 193 | Sungur C | *Ubaid 2-3/I* | 5700-4700 B.C.? | | | | | | | | | "No coherent plan can be detected for this building which was built of ***tauf*** about 70 cm. thick." (Jasim 1985: 156); "**The wall made of "tauf"** is irregularly about 70 cm thick." (Fuji 1981: 188) | Jasim, 1985:156; Fuji, 1981:188. |
| 194 | | *Ubaid* | 5700-4700 B.C.? | | | | | | | | | **"The buildings are built of "tauf"**, not of mud-bricks." | Fuji, 1981:191. |
| 195 | Tekhut | *Aratashen ?* | | General | | | | | | m | | «à Tekhut, dans la plaine de l'Ararat, les bâtiments, dont la base est toujours excavée dans le sol (de 0,6 à 1,1 m), ont des superstructures en briques crues parallélépipédiques (40 x 30 x 10 cm) (bât. II, XII), **(ou bien en pisé)** (pl. 19,3-4)." (Chataigner 1995: 59) | Chataigner, 1995:59. |
| 196 | | | | II, XII | 40 | 30 | 10 | 1,33 | 19,20 | | F | | Chataigner, 1995:59; Torosjan, 1976, pp. 23-27. |

| N° | Site | Culture/Level | Date | Building | Lenght (in cm) | Width (in cm) | Hight (in cm) | Ratio (L/w) | Weight estimated (in kg) | Technique | Shape or layout | Citation | Bibliographic references |
|---|---|---|---|---|---|---|---|---|---|---|---|---|---|
| 197 | Telul eth-Thalathat | *Umm Dabaghiyah-Sotto*/XV | 6500-6300 cal. B.C.* | General | | | | | | C | L | "No sherds of bricks were found since the construction **was not built with bricks** […]. Here in the cross section we saw **oblong lumps of clay** which seemed to be sun-dried bricks. Some of them were reedish brown and some of them were blueish gray. Their sizes are not uniform: Some are 40 cm in lenght and 8 cm in thickness and some are smaller than that. If they are sun-dried bricks, joint would be necessary for connecting them. But **no joint** was recognized among lumps. If this observation is right, the lumps must have been in such a condition that they could be connected with one another without joint. Tempered wet soil would fit to this condition. But, in this case, **it cannot keep the form of square cakes**. The only possibility, then, is that they used **oblong lumps of halfdried clay**, which cannot be called sun-dried bricks […]. If we see if from technological point of view, this method is forestep of the method of using sun-dried bricks, or it is a technological stage between the use of mud and the use of sun-dried […]. But the writer before reported his discovery of similar clay lumps at Tal-i Gap in Marv Dasht at the south-western part of Iranian Plateau." (Fukai *et al.* 1970: 15-16) | Fukai et al., 1970, pp. 15-16; Fukai et al., 1981:65; *Nishiaki et al., 2005:65, tab. 2. |
| 198 | Tilk Tepe | *Halaf II* | 5800-5550 B.C. | General | | | | | | C | und. | «**Il faut ajouter que le pisé**, associé à la pierre, était le matériau de construction utilisé à Tilki Tepe, sur les bords du lac de Van, à la même époque (phase récente de la culture d'Halaf)" | Chataigner, 1995:59. |
| 199 | Tell Turlu | *Halaf* | 6400-5400 B.C. | General | | | | | | C | und. | "La matière argileuse qui constituait les murs ne subsistait qu'exceptionnellement." | Breniquet, 1987:113. |

| N° | Site | Culture/Level | Date | Building | Lenght (in cm) | Width (in cm) | Hight (in cm) | Ratio (L/w) | Weight estimated (in kg) | Technique | Shape or layout | Citation | Bibliographic references |
|---|---|---|---|---|---|---|---|---|---|---|---|---|---|
| 200 | Yarim Tepe I | *Halaf* | 6170-5678 cal. B.C.*? | | 90 | 15 | | 6 | | m? | | | Munchaev *et* al. 1973: 6; *Merpert *et al.* 1976: 43. |
| 201 | Yarim Tepe II | *Halaf IA/VI* | 5640-4570 cal. B.C.* | *Tholos* 41 | | | | | | C | LC | "It was built of **clay layers**, 5 cm. thick on average, cemented with each other by a mortar of yellowish clay." | Merpert *et al.* 1976: 45; *Munchaev *et* al. 1981: 266. |
| 202 | | *Halaf IA* | 6100-5950 B.C. | 44A-44B | | | | | | C | LC | "The walls were made, in a conventional manner, of **clay layers**, held together by a stratum of grey-greenish loamy clay." | Merpert *et al.* 1977: 89. |
| 203 | | *Halaf II/VI* | 5800-5500 B.C. | *Tholos* 37 | | | | | | M | | «Alors que les autres techniques sont indifférement utilisées dès les premiers temps de l'époque de Halaf, **la brique crue moulée n'existe pas dans les premiers niveaux de Yarim Tépé II**. Son emploi est une des caractéristiques de l'architecture de basse Mésopotamie, mais ne semble avoir été introduite qu'à partir du Halaf moyen (**Yarim II, niveau VI, tholos 37**) dans le Nord et n'est pas limité à un type précis de bâtiment." | Breniquet 1992: 84. |

# Chapter 3

# Climatic conditions and technical choices in earthen architecture: the example of Late Neolithic Dikili Tash (Eastern Macedonia, Northern Greece)

Paul Bacoup
ED 112 - Université Paris I Panthéon Sorbonne; UMR 7041 ArScAn - Protohistoire égéenne
paul.bacoup@gmail.com, corresponding author

Sandra Prévost-Dermarkar
UMR 7041 ArScAn – Protohistoire égéenne
sandra.prevost@laposte.net

**Abstract**

The study of architectural remains conducted over the last twenty years on the tell of Dikili Tash in Greek Eastern Macedonia highlights the technical choices of prehistoric people during the Neolithic period (5000-4250 BCE). This article discusses the relations between the climatic conditions and these technical choices made by the builders. Thus, the presence of different vegetal tempers in the 'building earth' shows the integration of construction in the management of cereal crops. The privileged use of the technique of daub on joined posts for the walls requires abundant wooden resources. The study has also shown that some techniques have been chosen to isolate and protect the most fragile parts of the building against the variations in humidity and the action of water. However, these climatic conditions are not always decisive in the technical choices made, which are also explained by cultural factors.

**Keywords:** Environment, Earth construction, Aegean Neolithic, Wooden frameworks

**Résumé**

L'étude des vestiges architecturaux menée depuis une vingtaine d'années sur le tell de Dikili Tash en Macédoine orientale grecque met en évidence les choix techniques des populations durant le Néolithique (5000-4250 AEC). Cet article tente de souligner les relations entre les conditions climatiques et ces choix techniques. Ainsi, la présence de dégraissants végétaux dans la 'terre à bâtir' montre l'intégration de la construction dans la gestion des récoltes de céréales. L'utilisation privilégiée de la technique du torchis sur armature en poteaux jointifs pour les murs suppose des ressources en matière première arbustive abondantes. L'étude a aussi montré que certaines techniques ont été choisies pour isoler et protéger les parties les plus fragiles de la construction contre les variations d'humidité et l'action de l'eau. Cependant, ces conditions climatiques ne sont pas toujours déterminantes dans les choix techniques opérés, qui dépendent aussi de facteurs culturels.

**Mots-clés:** environnement, construction en terre, Néolithique egéen, armatures de bois

## 1 Introduction

The study of the relationship between climatic conditions and technical choices made by human societies in the past is currently a much-discussed topic. Research on environmental conditions and climatic changes has been increasing for decades in many domains, notably about their impacts on societies (Beniston 2005; Gang 2007; de Perthuis 2010; Stuckenberger 2010; Euzen et al. 2017). Technical studies and ethnographical surveys within present societies highlight a link between construction techniques and environmental constraints (Pilgram 1983): for example, the noticeably growing use of wood-frame constructions in Europe and even North America over the last two thousand years (Benoît and Paradis 2007). Approaching this question in Archaeology seems very important, yet easier said than done. Depending on the period, the area and the quality of the archaeological material at our disposal, answers are not systematically found. In this paper we discuss the relations - or the lack of them - between weather conditions and technical choices in earthen construction, based on the archaeological data acquired at the Late Neolithic site of Dikili Tash, Greece.

### *1.1 Presentation of the site and excavations*

The site of Dikili Tash is located in the south-eastern part of the plain of Drama in Eastern Macedonia (Northern Greece) (Figure 1). This site is a *tell*, which is an artificial and anthropic mound formed by a vertical accumulation of soil from the different occupation levels. These levels are mostly composed of burnt layers from the Neolithic and the Bronze Age. The mound is about 16 m high and covers a surface of 4 ha, making it one of the biggest mounds in the entire Aegean-Balkan world (Koukouli-Chryssanthaki and Rhomiopoulou 1992; Treuil 1992; Darcque and Tsirtsoni 2010).

The archaeological data place the Late Neolithic period of Dikili Tash between 5400 and 4250 BCE (Tsirtsoni 2016: 453-456). Our study is based on architectural remains discovered in two different sectors of the site, each one representing a different moment within the Late Neolithic period.

### *1.2 Presentation of the material*

The sector V was excavated during the second program of research, between 1986 and 1995. This sector extends over 145 m$^2$ on the south side of the mound (Figure 2). It corresponds to 5 levels of occupations dated between 5000 and 4800 BCE (Late Neolithic I). In some of these levels, architectural remains are well-preserved (Koukouli-Chryssanthaki and Treuil 2008: 7-14).

The excavation of sector 6 was also undertaken during the second research program (Darcque et al. 2007; Koukouli-Chryssanthaki et al. 1996) and during the third program (2008-2013) (Darcque 2013; Darcque et al. 2015). This sector covers an area of 565 m$^2$ on the south-eastern side of the mound. Five dwellings from the Late Neolithic II (ca. 4340-4250 BCE) were found. The architectural material discovered in sector 6 is being currently studied.

Figure 1. Map of Aegean area and location of Dikili Tash site.

Fires which destroyed the consecutive occupations were the main conservation factor of the earthen construction material. During the excavation, architectural remains had different aspects: heaps of fragments, scattered fragments and, to a lesser extent, fragments in connection, whether partially preserved in their last location before the fire or consequently collapsed. All fragments were analyzed, except those whose shapes were inconclusive, that had no surface or imprints, or were too small (Prévost-Dermarkar 2019: 9-11).

### 1.3 Methodological approach

Until the 1990s, although large quantities of burnt daub fragments were discovered on Neolithic mounds in the Aegean-Balkan world, these had not been collected, nor systematically recorded and even less studied. Architectural studies encountered various problems and challenges. The main problems were the identification, management and storage of thousands of fragments. The

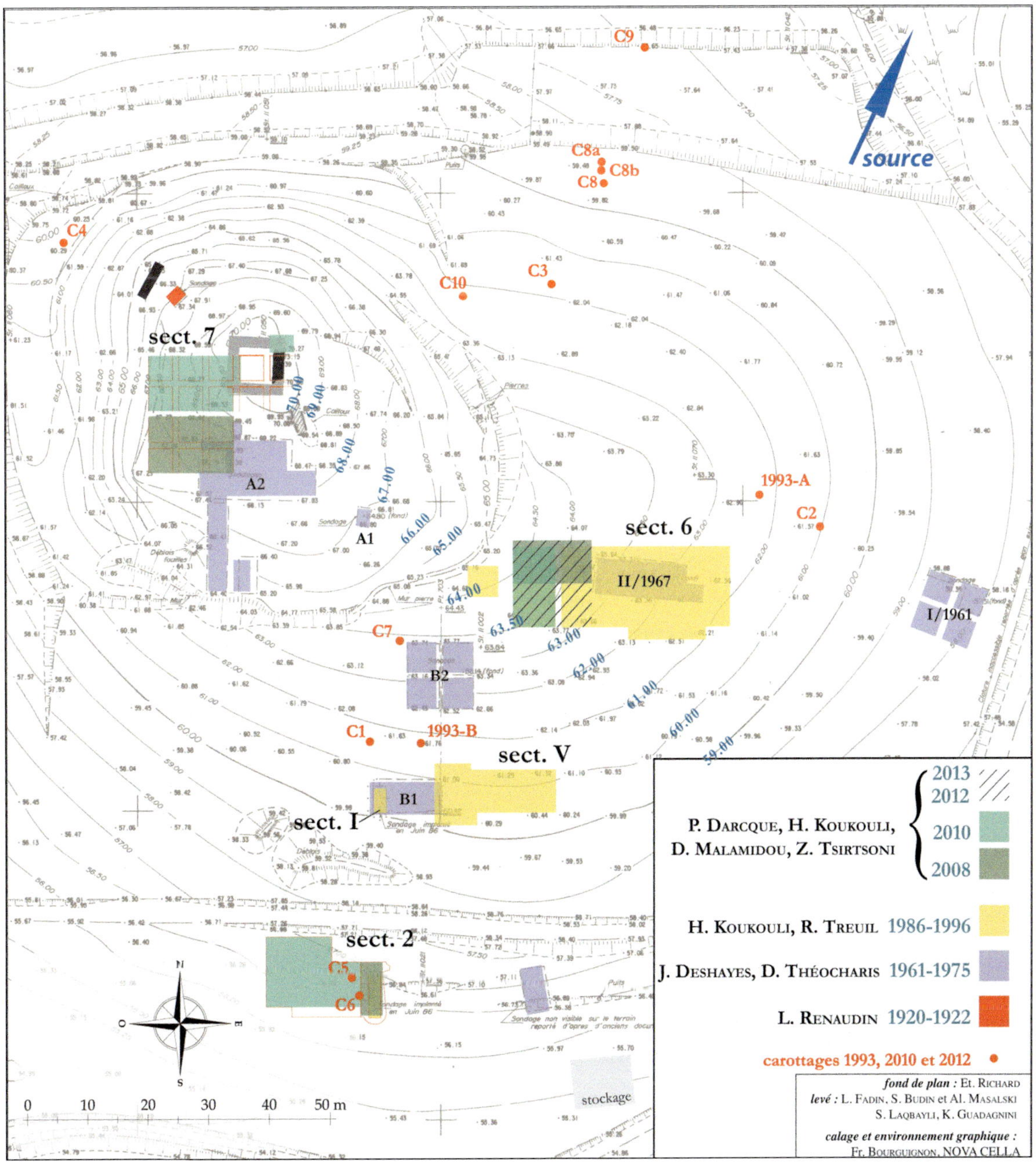


Figure 2. Topographical plan of the tell of Dikili Tash with location of excavation sectors until 2010.

objective of studying daub was to specify the technological know-how of Neolithic populations, to define the possible choices made by these populations as to how they used the resources available, to reconstruct the different operational sequences necessary for the construction of the houses, and thus to clarify the economic and social functioning of Neolithic villages (Prévost-Dermarkar 2019).

Building earth is a well-known material. Techno-lexical studies that have been taking place in the last forty years allow to identify the material with accuracy and to normalize the vocabulary – albeit in French –, and use it for description and study (de Chazelles and Poupet 1985; Dandrau 1997; de Chazelles and Klein 2003; Aurenche 2003; Aurenche 2004; Guillaud and Houben 2015). Starting in 1993 at Dikili Tash, a method to study architectural fragments of building earth was developed, based at first on a macroscopic morpho-technological description (Martinez 2001; Martinez et Prévost-Dermarkar 2003; Prévost-Dermarkar 2003; Perello and Prévost-Dermarkar 2018; Prévost-Dermarkar 2019). Then, in a second phase, interdisciplinary approaches were integrated. Micromorphological and physicochemical analyses, coupled with geomorphological research, were used to determine the morpho-pedological environment of the site (Dandrau 1997; Lespez 2008; Germain-Vallée et al. 2011). Archaeobotanical studies defined the vegetal tempers used in the building earth (Martinez 2001; Germain-Vallée et al. 2011). Experiments and ethnoarchaeological observations were used to verify technical hypotheses (Prévost-Dermarkar 2019) and traceological analyses on the archaeological tools to compare with the traces observed on the tools used in our experiments.

Furthermore, for heterogeneous wood-and-daub structures cases, both building earth fragments and timber have to be studied in concert. Initiated at the beginning of the project on sector V remains (Martinez 2001; Martinez and Prévost-Dermarkar 2003; Prévost-Dermarkar 2019) and then on the House 1 of sector 6 (Perello and Prévost-Dermarkar 2018), this technological analysis of construction wood continued in 2016 with the architectural study of the archaeological remains from Houses 3 and 4. Contrary to building earth, lexical approaches on wood used as construction material are not so common. A great synthesis by Volmer and Zimmermann (2012) is a model, which needs to be adapted to our data from Neolithic material. Based on previous studies made on the subject for the site of Dikili Tash (Martinez and Prévost-Dermarkar 2003; Prévost-Dermarkar 2003), as well as on glossaries known in other connected domains (Martinez 1996; Aurenche 2004), both a specific vocabulary and dimensional categories were established in order to avoid lexical ambiguity and also to optimize technical and statistical analysis of Neolithic construction wood (Bacoup 2018: 21-24).

This way, through the complementarity of the methods used in the current studies, knowledge was acquired on building earth composition and its preparation, on wood-frame structures, on surface finishing and on the building techniques of the architectural elements.

## 2 Climatic conditions and natural or farmed resources

### *2.1 Wood as an essential element in earth construction*

On the settlement of Dikili Tash, building earth is rarely used for 'homogeneous structures' (Aurenche 1981: 114-121) but rather in 'heterogeous structures' 'with wooden framework' where it is used as daub (Aurenche 1981: 127-128). The study of construction wood used for these frameworks is based for a great part on the imprints left by the wooden pieces on the daub fragments (Figure 3).

#### *2.1.1 Wood-framed construction: which species of wood were chosen*

There are different types of wooden framework associated with the different techniques of daub. These techniques of application allow filling empty spaces between the vegetal elements of the

Figure 3. Construction earth fragment. Imprints of joined posts from the framework of the eastern wall of House 4 (sector 6) (photo by P. Bacoup, EFA).

frame in order to insulate the structure, and also in order to solidify and strengthen the dwelling. The wooden armatures are organised in different manners involving different components. These can be sorted into five categories: (1) wattle-and-daub, (2) daub on joined posts, (3) daub on joined or bundled reeds/sticks, (4) wood-and-daub 'sandwich construction' and (5) daub on vegetal bed (Roodenberg 1995: 44-51; Martinez 1996: 35-38; Bacoup 2017: 67-68). Each of these framework types presents distinct technical features and provides information about wooden resources management and availability in the environment. They do not require the same quantity of wood, nor the same characteristics displayed by the trees chosen for lumber.

The conceptualisation of the wooden pieces by the builders enables them to define in advance the species of wood that they will need (genus, morphology, dimensions, etc.) (Alix and Brewster

2004). The selection is made based on two important criterions: the physicochemical properties of wood varieties and the morphological and dimensional features of trees.

Firstly, one of the advantages of building in daub on a wooden framework is that is does not require long and straight pieces. This way, a selection of the trunks based on their morphological characteristics can be ruled out. However, the selection of wooden architectural elements in accordance with dimensions is visible by the imprints. At Dikili Tash during the Late Neolithic period, dimensions were rather homogeneous. For example, ca. 60% of wooden pieces used as posts from the material of House 4 studied in 2016 and 2017 show a diameter ranging from 5 to 10 cm. In sector V, the study of a fallen wall block showed that for a similar technique, posts also had diameters from 5 to 9 cm (Prévost-Dermarkar 2019). The apparent standardization of diameters could have two different origins: the presence in the surrounding forest of trees that display trunks of similar dimensions, due to the growing conditions in dense primary forest for example, or it could be that Neolithic populations have chosen the trunks based on their diameter.

Secondly, the physicochemical properties are an important criterion for the selection of trees. This is true about lumber as well as for the other domains where wood was put to use (firewood, furniture, etc.). Early architecture treaties mention the most suitable species of wood founded on their innate qualities for construction purposes (Théophraste, Recherche sur les plantes, V; Vitruve, De architectura, II, 9, 5). However, to understand this choice in archaeology, it is necessary to know which genus were available and used in the context of our study (Bacoup 2018: 57-60).

*2.1.2 Wood-frameworks used at Dikili Tash*

On the site of Dikili Tash, among the five categories of frameworks known in general, three were identified over the different studies based on the architectural material: wattle-and-daub, daub on joined posts and to a lesser extent daub applied on a vegetal bed.

Wattle used as a vegetal armature is known for Dikili Tash, albeit sporadic (Figure 4). Indeed, this technique is less observed among the imprints, but it is visible on the dwelling remains from sector V (Prévost-Dermarkar 2019) as well as those from sector 6. So, wattle might have been preferred for particular architectural elements such as partition walls or gables, and not for main elements as perimeter walls. Ethnographic studies show us that the gables of Balkan dwellings often present different construction techniques, even different construction materials. In a similar fashion, the erection of partition walls in wattle-and-daub for dwellings that have perimeter walls made with different techniques and/or different materials is known during the Neolithic, notably for the Near East (Aurenche 1981: 143-145).

Within the environment near to the site, genuses used to make wattle – mainly viburnum (Viburnum sp.), cornelian cherry (Cornus mas), hazel (Corylus sp.) (Pétrequin 1991: 48) – are available during the Late Neolithic (Lespez et al,. 2000; Lespez 2008; Glais et al. 2016). These genuses were identified by anthracology in archaeological remains, especially in sector 6 dating from the Late Neolithic II (Malamidou et al. 2018). It is therefore difficult to assert that the motivation of the Late Neolithic populations at Dikili Tash for using wattle was solely based on environmental factors. It seems that this choice may be better understood through other factors: cultural, technical, etc.

While wattle frameworks are not common, perimeter walls of Late Neolithic dwellings at Dikili Tash were mainly erected by the application of daub on joined posts (Figure 5). The wooden pieces used for construction have a diameter generally ranging from 3 to 13 cm, including posts. They can be split, squared-off or used without any transformation. The framework is fixed by some crossties and oblique sticks (sector V see Martinez 2001; Prévost-Dermarkar 2019), that hold the standing posts together before the application of earth and that serve as mounts during the application of daub. Joined posts frameworks require a lot of wood. For example, 400 posts were necessary for

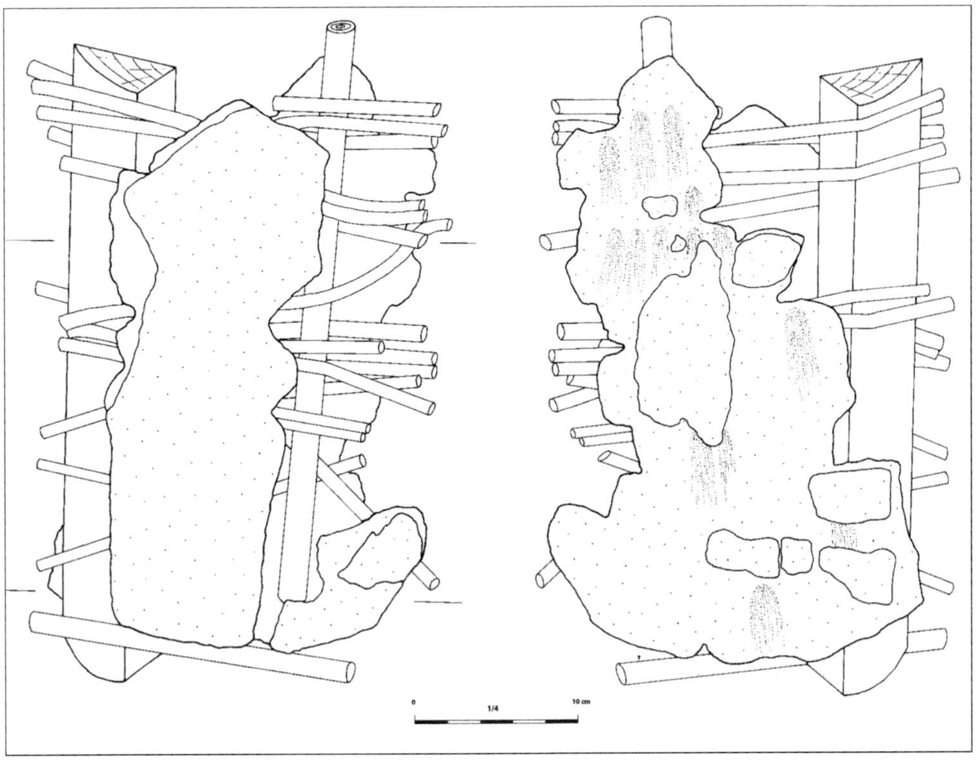

Figure 4. Technical drawing of construction earth fragment. Reconstitution of wattle-and-daub construction technique (drawing by P. Pugsley, EFA).

House 4, which meant that 120-130 trees had to be felled. Calculations brought the estimation up to 133 felled trees and discussing the topic with carpenters and loggers led to an estimate between 100-150. It is possible to understand this quantity by observing the diameters of the pieces. Indeed, it takes less than 20 minutes to fell trees with a diameter inferior to 13 cm using a stone-axe, while it can take up to one hour should the diameter be between 13 and 30 cm (Pétrequin 1991: 29; Mathieu et Meyer 1997: 234).

*2.1.3 Role of environment and climatic conditions on the wood technical choices*

At the end of the Early Neolithic and during the Middle and Late Neolithic (I and II), the plain of Drama knew favorable climatic conditions. During the Late Neolithic I, the forest environment was closing in. However, the fact that the oak groves have stayed at bay from the site has allowed riparian genuses to multiply, notably alder (Alnus sp.). According to the anthracological analyses of S. Thiebaut and M. Ntinou, oak (Quercus sp.) was mainly used for the making of the posts with a few occurrences of ash tree (Fraxinus sp.) and of juniper (Juniperus sp.). It is within the oak groves that these vegetal genuses are present. Around 4450 BC, during the Late Neolithic II, the environment close to site was intensively cleared. Consequently, the riparian species became scarcer and gave way to herbaceous areas and open spaces where anthropozoogeneous plants and cereals were cultivated, and livestock was bred. This change is mainly linked to human activities, which destroyed the biggest part of alder groves (Glais 2017).

In our case using the technique of daub on joined posts constitutes a true technical and architectural choice linked with the knowledge of physicochemical properties of wooden genuses. This choice is the result of a desire to build wall frameworks with posts made of oak, ash tree or juniper, even though other plants grew closer to the site, allowing the builders to use different techniques for their frameworks (wattle-and-daub) and/or other genuses (alder, willow (Salix sp.)). Ethnography shows us that choices ensue sometimes from exterior causes. However, it is difficult to explain the choices of wood species that were made by the Late Neolithic populations of Dikili Tash based solely

30 cm

Figure 5. Construction earth fragment. Imprints of joined posts and a crosstie from the framework of the eastern wall of House 4 (sector 6) (photo by P. Bacoup, EFA).

on an environmental determinism, 'perception of material and their properties' leads to take 'other [technical and cultural] factors' into account (Procopiou 2001: 241-243).

### 2.2 'Building earth' as the main material of earthen construction

#### 2.2.1 The basic components of 'building earth'

The mixture, called 'building earth' (in French 'terre à bâtir', see Aurenche 2003), used to cover these wood-frames of the walls and to build interior fittings (platforms, ovens, floors), consists of earth with more or less clay, water and tempering material (organic and/or mineral). This term is preferred to the term 'daub' which must be reserved for the technique of implementation of building earth on a wood-frames (Prévost-Dermarkar 2019).

In Dikili Tash, the observation of the prints left by the vegetal tempers in this mixture showed a difference in the nature and the proportions of the tempers used, depending on the kind of structure that the Neolithic craftsmen wanted to build (Malamidou et al. 2018: 66 and 69; Perello and Prévost-Dermarkar 2018).

#### 2.2.2 Selected vegetal tempers

Chopped straw

For all the walls of the dwellings in sector 6, the builders used chopped straw to prepare the building earth (Figure 6). This use is very frequently attested in archaeology (Bonnaire and Tengberg 2007) as well as in traditional earthen architecture. The morpho-technological study of architectural fragments (Prévost-Dermarkar, forthcoming) has shown that the size of the straw sprigs varies, as well as their quantity according to the layer of the wall (infra 3. 3. 2).

Chaff

The other vegetal temper known to Dikili Tash is chaff. In both sectors, it was used exclusively in the building earth used for all the interior fittings (oven vaults, platforms, floors, storage bins), as well as for the walls of the houses in sector V dating from the Late Neolithic I (Martinez 2001; Martinez and Prévost-Dermarkar 2003). Again, the amount of chaff varies, but this time, depending on the element built. The quantity is very important for the oven vaults (Figure 7) and less for the platforms (Prévost-Dermarkar 2003; Prévost-Dermarkar 2019).

Figure 6. Imprints of chopped straw inside a wall fragment of House 1 (sector 6) (photo by B. Perello, EFA).

The selection of the nature and quantity of the vegetal tempers used illustrates the high level of knowledge of the Neolithic builders. They knew how to meet the mechanical constraints of building earth while taking into account the quality of the earth chosen.

*2.2.3 Weather conditions encouraging cereal crops*

This intensive use of chaff and straw fits perfectly into the economic context highlighted by the geoarchaeological and palaeoenvironmental studies conducted over the last 25 years in the Drama plain by L. Lespez (Lespez 2008: 247-308 and p. 282, fig. 71) and A. Glais (Glais 2017: 484, fig. IV and p. 485, fig. V). They showed that with the Holocene climatic optimum (8300-4800 BCE), a hotter and more humid climate, combined with an absence of frost in winter, favored the development of the forest cover. During the Late Neolithic II, the landscape changed, impacted by anthropogenic activities: the cultivated fields expanded leading to the decline of tree species (supra 2.1.3). Recent studies have shown the intensification of cereal cultivation at Dikili Tash during this phase (Glais et al. 2017: 417). This picture was confirmed by the archaeobotanical studies carried out by S.M. Valamoti (2004 2015) which showed a wide variety of cultivated species, especially for cereals: einkorn wheat (Tritticum monococcum), barley (Hordeum vulgare) and emmer wheat (Tritticum dicoccum).

Figure 7. Imprints of chaff inside a fragment of vault's oven (sector V) (photo by P. Darcque, EFA).

The demonstration of the quasi-systematic use (with the exception of the mixture used for the floor of thermal structures) of vegetal tempers (straw and chaff) in the building earth sheds a new light on the Neolithic village organization. It highlights a reasoned management of crop residues with a storage obligation. It indicates that the *chaîne opératoire* in construction is integrated into the harvesting and processing of cereals. Moreover, that there was an increase in cereal culture at that time (Prévost-Dermarkar 2019).

## 3 Constraints of climatic conditions on building technical choices

### *3.1 Climate and earthen architecture*

Earthen architectures in general are vulnerable to climatic variations and foul weather, and so did the ones in Dikili Tash. In particular, they were undoubtedly put to a severe test by weather conditions, especially by a direct action of water and were extremely reacting to humidity variation. It is well known to traditional builders and modern engineers that moisture penetrates through permeability and capillarity, favored by the porosity of the earthen building material. The most critical sections of the building are the tops and bases of walls as well as openings (Guillaud and Houben 2015: 242). The water infiltrates the walls by the cracks (Figure 8) which often come at the points of contact between materials of different nature (earth/wood or earth/stone). They can be structural cracks or drying cracks (Guillaud and Houben 2015: 243). Both weaken the architectural element and make it more vulnerable to humidity, water or heat. The material can lose its cohesiveness and disintegrate.

### *3.2 Technical answers known in traditional architecture but not at Dikili Tash*

'The water resistance of an earth wall is above all dependent on the quality of the earth itself, its texture and structure, its porosity' (Guillaud and Houben 2015: 326). The contribution of a natural stabilizer (oils, sap or urine) can improve this resistance (Vissac et al. 2017), but they are, for the moment, impossible to bring out by analyses of archaeological fragments.

'For regions of temperate climate, (...) the earth walls resist the erosion of bad weather provided they are raised on good foundations and good substructures, and that they are protected in their upper part' (Guillaud and Houben 2015: 326) (Figure 9). However, no drainage system to prevent water stagnation near the buildings was identified during excavations at Dikili Tash, nor any stone foundation at the base of the walls to limit capillary rise. The few remains of wall bases preserved in place show that the joined posts of the framework were sunk directly into the ground. As for roofing, in this region during the Late Neolithic period we generally infer the presence of thatched gable roofs with ample overhang (Treuil 2008: 69-71), which act as efficient rainwater protection in traditional architecture (Efstratiou 2002: 369). But there is no archaeological evidence to support this hypothesis so far.

Figure 8. Cracks around an opening in a traditional house (Kovačevo, Bulgaria) (photo by S. Prévost-Dermarkar).

### *3.3 Technical answers at Dikili Tash*

As the traditional architecture shows, when the construction is not well designed to withstand bad weather (lack of foundations for the walls, inadequate roofing), it becomes necessary to apply a protective coating against wind and rain (Guillaud and Houben 2015: 327). Earth plasters form a layer which will be subjected to erosion that will require regular repairs.

#### *3.3.1 Multiple layers of daub*

On the walls of the houses in the two sectors of Dikili Tash, we observed that

Figure 9. A traditional house with a stone foundation (Kovačevo, Bulgaria) (photo by S. Prévost-Dermarkar).

several layers of daub were applied on the joined posts framework (Figure 10). Their thickness can range from a minimum of 5 cm to a maximum of 11 cm (Perello and Prévost-Dermarkar 2018: 188). These layers are identical to one another from the point of view of the nature of their components: earth and vegetal temper. Thus, earth with more or less clay always shows the presence of anthropogenic elements (such as sherds, micro-fragments of bone, charcoals, seeds, nodules of building earth), suggesting the use of earth coming from the tell. These macroscopic observations were confirmed by the physicochemical, micromorphological and geomorphological analyses carried out in the 2000s (Lespez 2008, Prévost-Dermarkar 2019). However, it is difficult to say whether these multiple layers of daub correspond to the initial phase of construction of the wall to give it the desired thickness or if some are subsequent restoration layers, related to degradation.

*3.3.2 Plasters of varying thicknesses*

However, layers of multiple plasters, varying thicknesses (Figure 11) and different compositions have often been observed on the remains of the best preserved walls of the houses in sector V (Late Neolithic I) (Prévost-Dermarkar 2019: 31) and sector 6 (Late Neolithic II) (Perello and Prévost-Dermarkar 2018: 189).

Figure 10. Varied adding layers on archeological fragment of wall (photo by B. Perello, EFA).

Different compositions of plasters

On some of the very large fragments preserved (about one meter long by 40 cm wide and 20 cm thick, Perello and Prévost-Dermarkar 2018: 187, fig.4), these plasters are distinguished from the internal layer of building earth by the quantity and type of vegetal temper. Especially for the walls of house 1 in sector 6, we observed that in the inner layers of building earth, the fibers can be very large (5 mm in diameter) and very numerous (Figure 12), whereas the layers of plaster superimposed over them, are increasingly fewer and thinner (up to less than 1 mm). The composition of the last layer perhaps indicates the desire for a smoother final result (Perello and Prévost-Dermarkar 2018: 187, fig.5).

Figure 11. Plasters of varying thicknesses on archeological fragment of wall (photo by B. Perello, EFA).

Figure 12. Plaster layer with very numerous prints of fibers on archeological fragment of wall (photo by S. Prévost-Dermarkar, EFA).

Different colors of plasters

Some fragments have been found with a colored or whitish layer (Figure 13) which suggest the presence of a topcoat, sometimes as thin as a simple wash (Prévost-Dermarkar, forthcoming). It is very likely that some of these layers have been applied merely to protect or maintain the wall. They may also have contributed to the improvement of thermal comfort or may have had an aesthetic purpose or both.

### *3.3.3 Stone substructures of thermal structures*

Stone foundations that do not exist for the bases of the walls are, however, recurrent for domestic ovens, found to be well preserved in dwellings in both sectors, whether at ground level or raised (Germain-Vallée et al. 2011; Prévost-Dermarkar 2003; Prévost-Dermarkar 2019). In most cases, this base is at least 10 cm thick. In the case of the oven 614 in sector V (Figure 14), it reaches 40 cm with a succession of stone layers of different sizes (Darcque et al. forthcoming). This careful stone foundation undoubtedly responds to a need for a better insulation of the thermal structure against moisture rising by capillarity. Moreover, it reinforces thermic properties of the structure which has to store a maximum of heat first to restitute it in a second time (Papadopoulou and Prévost-Dermarkar 2007).

## 4 Conclusion

The study of architectural remains of Dikili Tash site illustrates the difficulty to understand the relationship between climatic conditions and technical choices of prehistoric men. The environment of the site during the Late Neolithic, as reconstructed by geoarchaeological and paleoenvironmental analyses, is favorable to earth-and-wood construction with wooden and farmed

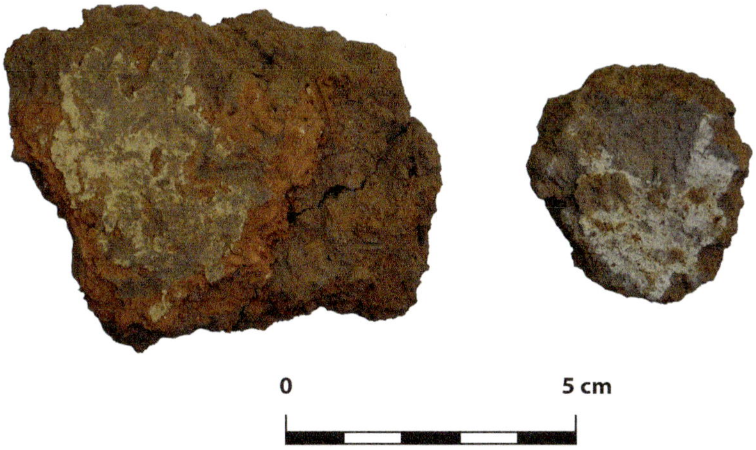


Figure 13. Plaster layers of different colors on archeological fragments of wall (photo by S. Prévost-Dermarkar, EFA).

resources in abundance. Thus, the analysis of 'building earth' composition highlights the use of chaff and straw that are by-product of cereal processing, suggesting the integration of construction in the plant management process.

Nevertheless, the architectural study has also shown that the chosen materials and techniques were not necessarily determined only by environmental conditions or available resources. The original joined posts building technique for the walls, rather than wattle-and-daub, during the two phases of the Late Neolithic, is motivated by technical and cultural reasons rather than by the availability of certain wood resources. However, it underlines the durability of this architectural tradition and the transmission of the technical know-how that was put to use over a thousand years.

To respond to climatic constraints, our study brings to light other technical choices as the use of plasters throughout the Late Neolithic. The absence of stone foundations for the walls can be explained by the building technique of joined posts. Regarding drainage systems of the dwellings, the specific taphonomical conditions of the *tell* prevents us from finding them: successive archaeological layers are often disturbed and the material was often reused.

Figure 14. Different layers of stones for the substructure of oven 614 (photo by P. Darcque, EFA).

## Bibliography


Alix, C. and Brewster, K. (2004) 'Not all Driftwood is Created Equal: Wood Use and Value along the Yukon and Kuskokwim Rivers, Alaska', *Alaska Journal of Anthropology*, 2(1): 48-65.

Aurenche, O. (1981) *La maison orientale. L'architecture du Proche Orient ancien des origines au milieu du quatrième millénaire. Tome 1. Texte.* Paris: Librairie orientaliste Paul Geuther S.A.

Aurenche, O. (2003) 'Propositions de terminologie pour les modalités de mise en œuvre de la terre comme matériau de construction' in Chazelles C.A. de and Klein, A. (eds) *Échanges transdisciplinaires sur les constructions en terre crue. Actes de la table ronde de Montpellier 17-18 novembre 2001.* Montpellier: Éd. de l'Espérou: 279-282.

Aurenche, O. (ed.) (2004) *Dictionnaire illustré multilingue de l'architecture du Proche Orient ancien.* Lyon: Maison de l'Orient et de la Méditerranée.

Bacoup, P. (2017) Les techniques liées au travail du bois de construction. Le monde égéen et son pourtour balkanique de 6500 à 4000 avant notre ère. Volume 1. Texte. Mémoire de Master 1. Paris: Université Paris I (Panthéon-Sorbonne).

Bacoup, P. (2018) Le bois de construction : un témoin environnemental, technique et architectural des sociétés du passé. Un exemple régional en Bulgarie du Nord au Chalcolithique. Mémoire de Master 2. Paris: Université Paris I (Panthéon-Sorbonne).

Beniston, M. (2005) 'Changement climatique et impacts possibles dans la région alpine', *Revue de géographie alpine*, 93(2): 13-24.

Benoit, Y. and Paradis, T. (2007) *Construction de maisons à ossature bois.* Paris: Eyrolles.

Bonnaire, E. and Tengberg, M. (2007) 'Éléments pour l'identification des empreintes végétales dans la terre à bâtir' in Van Doosselaere, B. and Oberweiler, C. (eds), *Dégraissants organiques: identifications, nomenclatures et référentiels. Actes de la table-ronde. Cahier des thèmes transversaux ArScAn 7*: 81-85.

Dandrau, A. (1997) La construction en terre dans le monde égéen protohistorique. Thèse de doctorat. Paris: Université Paris 1 (Panthéon-Sorbonne).

Darcque, P. (2013) 'Dikili Tash, un village néolithique dans le Nord de la Grèce', *CRAI*, I (janvier-mars): 51-73.

Darcque, P. Koukouli-Chryssanthaki, H., Malamidou, D., Treuil, R. and Tsirtsoni, Z. (2007) 'Recent researches at the Neolithic settlement of Dikili Tash, Eastern Macedonia, Greece: an overview' in Todorova, H., Stefanocich, M. and Ivanov, G. (eds), *The Struma/Strymon river valley in prehistory: proceedings of the International Symposium Strymon Praehistoricus, Kjustendil-Blagoevgrad (Bulgaria), Serres-Amphipolis (Greece).* Sofia: Gerda-Henkel-Stiftung: 247-256.

Darcque, P. Koukouli-Chryssanthaki, H., Malamidou, D., Treuil, R. and Tsirtsoni, Z. (forthcoming) Dikili Tash, village préhistorique de Macédoine orientale, Histoire d'un tell: les recherches 1986-2016, vol. II, 2.

Darcque, P. and Tsirtsoni, Z. (2010) 'Evidence from Dikili Tash (Eastern Macedonia, Greece) and the tell issue' in Hansen, S. (ed.), *Leben auf dem Tell als soziale Praxis, Beiträge des Internationalen Symposiums in Berlin vom 26.-27. Februar 2007.* Bonn: Dr. Rudolf Habelt GmbH: 55-69.

Darcque, P. Tsirtsoni, Z., Koukouli-Chryssanthaki, H. and Malamidou, D. (2015) 'New Insights to the Copper Age Economy and Chronology at the Tell Settlement of Dikili Tash (Northern Greece)' in Hansen, S., Raczky P., Anders, A. and Reingruber, A. (eds) *Neolithic and Copper Age between the Carpathians and the Aegean Sea: chronologies and technologies from the 6th to the 4th millennium BCE; International workshop Budapest 2012.* Bonn: Habelt: 403-417.

de Chazelles, C.-A. and Klein, A. (eds) (2003) *Échanges transdisciplinaires sur les constructions en terre crue : Actes de la table-ronde de Montpellier, 17-18 novembre 2001.* Montpellier: Éd. de l'Espérou.

de Chazelles, C.-A. and Poupet, P. (1985) 'La fouille des structures en terre crue : définitions et difficultés', *Aquitania*, 3: 149-160.

de Perthuis, C. (2010) *Et pour quelques degrés de plus: changement climatique : incertitudes et choix économiques.* Paris: Pearson education France.

Efstratiou, N. (2002) *Εθνοαρχαιολογικές αναζητήσεις στα Πομακοχώρια της Ροδόπης.* Thessaloniki: Εκδόσεις Βάνιας.

Euzen, A., Laville, B. and Thiébault, S. (2017) *L'adaptation au changement climatique: une question de sociétés.* Paris: CNRS éditions.

Gang, H. (2007) 'La société chinoise face au changement climatique', *Perspectives chinoises*, 98(1): 79-85.

Germain-Vallée, C., Prévost-Dermarkar, S. and Lespez, L. (2011) 'Stratégie de prélèvement et de mise en oeuvre de la 'terre à bâtir' des structures de combustion néolithique du site de Dikili Tash (Grèce) à partir d'une étude micromorphologique', *Archéosciences*, 35: 41-63.

Glais, A. (2017) Interactions Sociétés-Environnement en Macédoine orientale (Grèce du Nord) depuis le début de l'Holocène. Approche multiscalaire et paléoenvironnementale. Thèse de doctorat. Caen: Université de Caen Normandie.

Glais, A., López-Sáez, J.A., Lespez, L. and Davidson, R. (2016) 'Climate and human-environment relationships on the edge of the Tenaghi-Philippon marsh (Northern Greece) during the Neolithization process', *Quaternary International*, 403: 237-250.

Glais, A., Lespez, L., Vannière, B. and López-Sáez, J.A. (2017) 'Human-shaped landscape history in NE Greece. A palaeoenvironmental perspective', *Journal of Archaeological Science: Reports*, 15: 405-422.

Guillaud, H. and Houben, H. (2015) *Traité de construction en terre.* CRATerre (4th ed.). Marseille: Parenthèses.

Koukouli-Chryssanthaki, H. and Romiopoulou A. (1992) 'Οι ανασκαφές στον ελληνικό τομέα του προϊστορικού οικισμού Ντικιλί Τας (1961-1967)' in *Διεθνές Συνέδριο για την Αρχαία Θεσσαλία στη μνήμη του Δ.Ρ.* Θεοχάρη, Βόλος 1987: 226-248.

Koukouli-Chryssanthaki, H., Treuil, R. and Malamidou, D. (1996) Προϊστορικός οικισμός Φιλίππων Ντικιλί Τας: δέκα χρόνια ανασκαφικής έρευνας, *AEMth* 10: 681-704.

Koukouli-Chryssanthaki, H. and Treuil, R. (2008) 'Le programme de recherches' in Koukouli-Chryssanthaki, H. and Treuil, R. (eds) *Dikili Tash, village préhistorique de Macédoine orientale: recherches franco-helléniques dirigées par la Société Archéologique d'Athènes et l'École Française d'Athènes; (1986 - 2001).* Athens: Bibliothèque de la Société Archéologique d'Athènes, 254: 3-19.

Lespez, L. (2008) 'L'évolution des paysages du Néolithique à la période ottomane dans la plaine de Philippe-Drama', in Koukouli-Chryssanthaki, H. and Treuil, R. (eds) *Dikili Tash, village préhistorique de Macédoine orientale: recherches franco-helléniques dirigées par la Société Archéologique d'Athènes et l'École Française d'Athènes; (1986-2001).* Athens: Bibliothèque de la Société Archéologique d'Athènes, 254: 21-394.

Lespez, L., Dalongeville, R., Noirel-Schtz, C., Suc, J.P., Koukouli-Chryssanthaki, H. and Treuil, R. (2000) 'Les paléoenvironnements du site préhistorique de Dikili Tash (Macédoine orientale, Grèce)', *Bulletin de correspondance hellénique*, 124(2): 413-434.

Malamidou, D., Ntinou, M., Valamoti, S.M., Tsirtsoni, Z., Koukouli-Chryssanthaki, H. and Darcque, P. (2018) 'An Investigation of Neolithic Settlement Pattern and Plant Exploitation at Dikili Tash: Reconsidering Old and New Data from the Late 5th Millennium B.C. Settlement' in Sarris, A., Kalogiropoulou, E., Kalayci, T. and Karimali, E. (eds) *Communities, Landscapes, and Interaction in Neolithic Greece.* New York City: Berghahn Books Incorporated: 60-80.

Martinez, S. (1996) Les techniques de construction au Néolithique dans la Grèce du Nord et les Balkans. Mémoire de DEA. Paris: Université Paris I (Panthéon-Sorbonne).

Martinez, S. (2001) 'A New Look at House Construction Techniques. Current Research at Dikili Tash, Neolithic Site of Eastern Macedonia', *AEMTh*, 13: 63-68.

Martinez, S. and Prévost-Dermarkar, S. (2003) 'Les techniques de construction de l'habitat en terre du site néolithique de Dikili Tash (Macédoine orientale, Grèce)', *Cahier des Thèmes transversaux, ArScAn*, IV: 147-150.

Mathieu, J.R. and Meyer, D.A. (1997) 'Comparing Axe Heads of Stone, Bronze, and Steel: Studies in Experimental Archaeology', *Journal of Field Archaeology*, 24(3): 333-351.

Papadopoulou, E. and Prévost-Dermarkar, S. (2007) 'Il n'y a pas de cuisine sans feu. Une approche des techniques culinaires au Néolithique et à l'Âge du Bronze Ancien à travers les structures de combustion en Grèce du Nord' in Mee, C. and Renard, J. (eds), *Cooking up the Past, Food and culinary practices in the Neolithic and Bronze Age Aegean.* Oxford: Oxbow Books: 123-135.

Perello, B. and Prévost-Dermarkar, S. (2018) 'L'architecture domestique en torchis du site de Dikili Tash : nouvelles données de la maison 1 du secteur 6 (Ve millénaire av. n. è.)'in de Chazelles, C.-A. and Klein, A. (eds) *4èmes Échanges transdisciplinaires sur les constructions en terre crue : Actes de la table-ronde de Lattes, du 23 au 25 novembre 2016.* Montpellier: Éd. de l'Espérou: 185-190

Pétrequin, P. (ed.) (1991) *Construire une maison 3000 ans avant J.-C. Le lac de Chalain au Néolithique.* Paris: Errance.

Pilgram, T. (1983) 'Environmental change and timber selection in huambisa Jivaro house construction', *Journal d'agriculture traditionnelle et de botanique appliquée*, 30(2): 139-147.

Prévost-Dermarkar, S. (2003) 'Les fours néolithiques de Dikili Tash (Macédoine, Grèce) : une approche expérimentale des techniques de construction des voûtes en terre à bâtir' in Frère-Sautot, M.C. (ed.) *Le feu domestique et ses structures au Néolithique et aux Âges des métaux. Actes du colloque de Bourg-en-Bresses et Beaune, 7 et 8 octobre 2000.* Montagnac: M. Mergoil: 215-223.

Prévost-Dermarkar, S. (2019) 'Bâtir en terre au Néolithique : approche morpho-technologique des vestiges architecturaux de Dikili Tash', *Bulletin de correspondance hellénique*, 143(1): 1-61.

Procopiou, H. (2011) 'IV – La reconstitution des mentalités. 2. Choix et décision, sens et émotions: le début des études de perception' in Treuil, R. (ed.), *L'archéologie cognitive: techniques, modes de communication, mentalités.* Paris: Maison des sciences de l'homme: 233-255.

Roodenberg, J.J. (ed.) (1995) *The Ilıpınar excavations. 1: Five seasons of fieldwork in NW Anatolia 1987 - 91.* Istanbul: Nederlands Historisch-Archeologisch Instituut.

Stuckenberger, A.N. (2010) 'Les Inuit et le changement climatique', *Études/Inuit/Studies*, 34(1): 5-19.

Treuil, R. (ed.) (1992) *Dikili Tash: Village préhistorique de Macédoine orientale I. Vol. 1: Fouilles de Jean Deshayes (1961-1975).* Athens: École Française d'Athènes.

Treuil, R. (2008) 'Le Paléolithique et le Néolithique. L'évolution de la civilisation' in Treuil, R., Darcque, P. Poursat, J.-C. and Touchais, G. (eds) *Les civilisations égéennes du Néolithique et de l'Âge du bronze.* Paris: 2nd ed:U.F: 63-82.

Tsirtsoni, Z. (2016) 'Concluding Remarks' in Tsirtsoni, Z. (ed.), *The Human Face of Radiocarbon. Reassessing Chronology in prehistoric Greece and Bulgaria 5000-3000 cal BC.* Lyon: Maison de l'Orient et de la Méditerranée: 453-464.

Valamoti, S.M. (2004) *Plants and people in Late Neolithic and Early Bronze Age northern Greece: an archaeological investigation.* Oxford: British Archaeological Reports, Series 1258.

Valamoti, S.M. (2015) 'Harvesting the 'wild'? Exploring the context of fruit and nut exploitation at Neolithic Dikili Tash, with special reference to wine', *Vegetation History and Archaeobotany*, 24: 35-46.

Vissac, A., Bourges, A., Gandreau, D., Anger, R. and Fontaine, L. (2017) *Argiles & biopolymères. Les stabilisants naturels pour la construction en terre.* Villefontaine: CRAterre éditions.

Volmer, L. and Zimmermann, W.H. (eds) (2012) *Glossar zum prähistorischen und historischen Holzbau: Französisch, Englisch, Niederländisch, Deutsch, Dänisch, Norwegisch, Schwedisch, Polnisch und Tschechisch. Glossary of prehistoric and historic timber buildings: French, English, Dutch, German, Danish, Norwegian, Swedish, Polish and Czech.* Rahden: Verlag Marie Leidorf.

# Chapter 4

# Earth construction procedures in the Arslantepe domestic architecture from the Late Chalcolithic to the Early Bronze Age

Giovanna Liberotti
Italian Archaeological Expedition in Eastern Anatolia
giovanna.liberotti@gmail.com

**Abstract**

For more than 12000 years, people have extensively used earthen materials to build living spaces. Many communities in different environments still possess in-depth knowledge of earthen materials and construction techniques transmitted generation by generation through millennia. Sun-dried mudbricks have been and are still used today as building material in monumental and domestic architecture at the archaeological site of Arslantepe, within the Malatya region in Eastern Turkey, 15 km southwest from the Euphrates River. More than 60 years of extensive excavation documented a long occupation starting at least with the Halaf period (6th millennium BCE) up to the Byzantine period. This paper shows how laboratory analyses on samples of earthen building material (i.e. x-ray fluorescence, macro-porosity and mechanical tests) may add information to the study of the ancient technology. The sampled structures have also been investigated from an architectural point of view in order to better understand their original context.

**Keywords.** Earthen Architecture, Near Eastern Archaeology, Archaeometry

**Résumé**

Pendant plus de 12000 ans, les gens ont largement utilisé des matériaux de terre pour créer des espaces où vivre. De nombreuses communautés possèdent encore une connaissance approfondie des matériaux en terre et des techniques de construction transmises de génération en génération. Les briques en terre crue séchées au soleil ont été et sont encore utilisées aujourd'hui dans le site archéologique d'Arslantepe (Malatya), dans l'est de la Turquie. Plus de 60 ans de fouilles intensives ont documenté une longue occupation commençant au moins de la période de Halaf (6ème millénium AEC) jusqu'à la période byzantine. Cet article montre comment les analyses de laboratoire sur des échantillons de matériaux de construction en terre (fluorescence des rayons X, macroporosité et tests mécaniques) peuvent ajouter des informations à l'étude de l'ancienne technologie. Les structures échantillonnées ont également été étudiées d'un point de vue architectural afin de mieux comprendre leur contexte d'origine.

**Mots-clés:** architecture en terre, archéologie du Proche-Orient, archéométrie

## 1 Introduction

The tradition of building with earth spans the history of human beings and covers a large geographical area, with a variety of typologies that reflect the cultural identity of the people who use it and their geographical and technical environments. The high level of technical variability, which characterizes earthen architecture, has generated a relevant heritage of orally transmitted empirical knowledge that is nowadays subjected to relegation and loss. Until now, scientific research has mostly approached in a descriptive way the technological aspects and social requirements relating to building with earth in Pre-and Protohistoric societies, disregarding the importance of understanding the fabrication procedures as well as the engineering and architectural know-how. Together with Egypt (Emery 2011) and the Near East (Homsher 2012; Sauvage 2011), Anatolia (Love 2013; Stevanovic 2003) is where the majority of the research and analytical work on prehistoric earthen construction materials come from, as these materials are often better preserved. The technical issues investigated through soil analysis that seem to have dominated the literature are the use of tempering in the clay paste (Forget et al. 2015), the properties of various types of sediment (Coffman et al. 1990) and the selection of raw materials to be used in earthen architecture (French 1984; Kemp 2000). It is now common knowledge that clay deposits, which apparently have the same basic features, are formed by lithological components varying greatly from one deposit to another and, to a lesser extent, within the same deposit (Nodarou, Frederick and Hein 2008). Variations in the composition also determine variations in the quality of clay. Depending on its physical and chemical characteristics, clay will adapt more or less appropriately to different uses.

Sun-dried mudbricks, together with stone and wood, were the building materials used for millennia in monumental and domestic architecture at the archaeological site of Arslantepe (Malatya), in Anatolia, Eastern Turkey. Almost sixty years of systematic extensive archaeological excavation allowed intensive observation and mapping of several superimposed settlements, providing the researchers with a certain discernment in the evolution/devolution of otherwise invisible architectural patterns. One of the aims of this work is to find out if, once attested that different kinds of clay do have a direct influence on the performance of earthen building materials, there is or there is not a precise technological choice behind this selection. Well-preserved mudbricks relevant to different settlement periods from Late Chalcolithic to Early Late Bronze Age have been examined and compared using a combination of laboratory analytical techniques and architectural analysis, in order to identify the component variability and the variation in performance as well as the technology of manufacture and the potential use of specific recipes. The connection between the architectural and archaeometric analysis of the sampled buildings suggests a certain degree of standardization in the recipes and manufacturing process of mudbricks, notwithstanding that at the turn of the second quarter of the third millennium some abrupt transitions occurred in the development of the communities that were living in Arslantepe.

## 2 Setting the scene: geo-morphological setup and historical framework

The artificial mound of Arslantepe is located 910 m a.s.l. in the Malatya plain, eastern Anatolia (Turkey), north of the Antitaurus Mountains and about 15 km southwest from the Euphrates River (Figure 1). It is 30 m high and about 4 ha in size and has a long occupation starting at least with the Halaf period in the 6th millennium BCE up to the Neo-Hittite period. Later, Roman and Byzantine occupation is also known on the site (Palmieri 1973; Puglisi and Meriggi 1964; Schneider Equini 1970). The mound (tepe in Turkish) consists essentially of the decay material, mainly mudbricks and stones, from the structures themselves, accumulated by every successive settlement over the millennia. The fertile alluvial soils and easily accessible aquifers of the Malatya plain still today ensure high agricultural potential. Despite drastic forest reduction over the millennia, recent studies on the paleo-climate and the determination of wood samples from the in situ burnt remains of poles (Dreibrodt et al. 2014; Marcolongo and Palmieri 1983; Sadori and Masi 2012) demonstrated that the duality between arid climate of the steppes and abundance of natural springs characterized the area also in the past. Stable carbon isotope analyses of plant remains (Arıkan, Balossi and Masi 2016; Baneschi et al. 2012; Masi et al. 2014) revealed that in the late fourth and third millennium BCE the climate was more humid than it is today. Woody species of the Malatya plain mainly consisted of deciduous and evergreen broadleaved plants such as oaks and rosaceae, and of hygrophilous taxa such as poplars and willows (Kaya and Raynal 2001). From a geological point of view, Arslantepe lies on Miocene lake sediments rich in calcareous clays, limestone and sandstone (Palmieri 1978). Northeast of the site is the Middle Miocene Orduzu volcanic suite (Aksoy, Türkmen and Turan 2005), which is part of the Neogene collision-related Yamadağ volcanism of Eastern Anatolia and includes calc-alkaline rhyolites, trachyandesites, basaltic trachyandesites and quartz-micromonzonites (Önal 2008; Önal et al. 2008). Further east, the Late Cretaceous calc-alkaline volcanic and intrusive rocks of the Baskil magmatics complex (Yazgan and Mason 1988) and the Maastrichtian to Early Eocene Yüksekova/Elazığ complex consist of a lower intact ophiolitic rock assemblage and an upper calc-alkaline unit intermediate to felsic lavas and associated sedimentary rocks (Yiğitbaş and Yılmaz 1996). To the south, the Malatya metamorphic units composed of Carboniferous to Triassic meta-carbonate rocks, mica schists, phyllites, slates, meta-clastic rocks and meta-cherts mainly ascribable to a greenschist facies (Bozkaya et al. 2007; Robertson et al. 2006) dominate the surroundings. This scenario is extraordinarily favourable to the exploitation of raw material such as clay for sun-dried mudbrick manufacture and stone for wall foundations and paving. Clay and stone can be easily extracted in the proximity of the settlement and/or production sites so that to reduce time and cost of acquisition and transportation (Arnold 2000; Homsher 2012: 18; Pritchard 1985: 35; Rosen 1986).

In 1961, the Italian Archaeological Expedition in Eastern Anatolia of the Sapienza University of Rome resumed the French excavation carried out at Arslantepe during the 1930s and 1940s documenting the Neo-Hittite and Hittite levels in the northernmost area of the mound (Alvaro 2012). The Italian team, nowadays directed by Prof. Marcella Frangipane, extended the excavation to a larger part of the mound bringing to light the evidence of a centralized proto-urban system of Syro-Mesopotamian influence (Frangipane 2012a) dating back at least to the mid fourth millennium BCE (3600-3400 BCE, Period VII) and characterized by elite residences, a large temple and domestic quarters. An impressive set of interconnected and functionally differentiated buildings discovered in a direct sequence above the period VII structures constitutes the first known example of 'Palace' in the Near East (3400-3100 BCE, Period VI A), where a wide range of cultic, centralizing and redistributive activities took place (Frangipane 1997 2001 2012a; Frangipane and Palmieri 1983). At

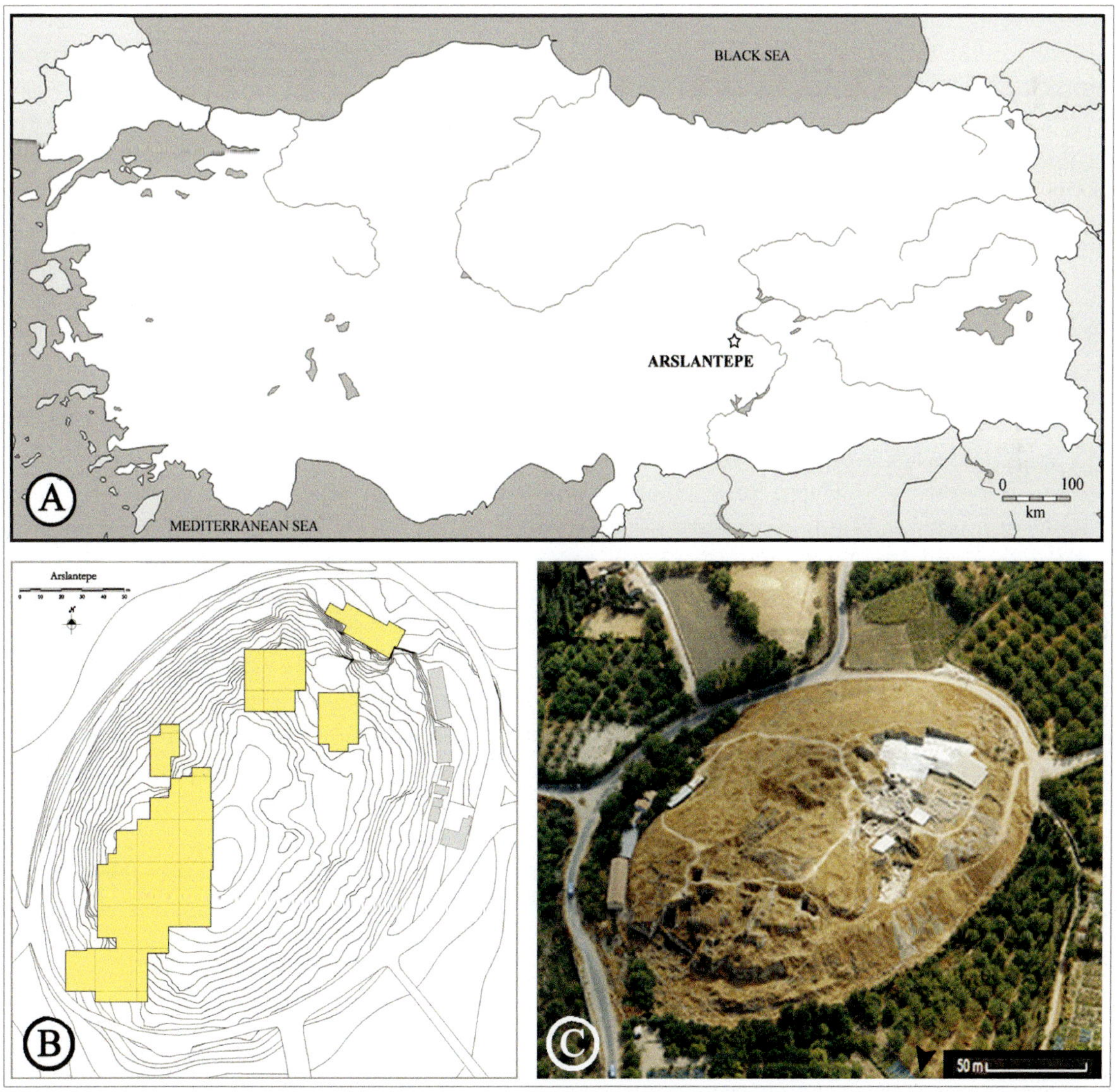


Figure 1. A) Map of Turkey with the localization of Arslantepe; B) Plan view of Arslantepe with the excavated areas in yellow colour; C) Eye bird view of Arslantepe - photo: F.O Durgut, © DAI.

the back of this monumental complex of buildings, elite residences were unearthed (Frangipane 2010). The excellent state of preservation, the more than fifty years of excavation and conservation, and the astonishing finds from these structures make them a unique case and source of knowledge on early state formation (Frangipane 1997, 2009). At the end of the fourth millennium, one or more fires permanently ravaged the palatial complex and the system collapsed. Agriculturalists and pastoral nomads with strong Transcaucasian connections settled in the vast area where the collapsed buildings were (3100-3000 BCE, Period VI B1), without completely replacing the previous cultural substratum but coexisting with it (Alvaro and Palumbi 2014; Palumbi 2012). Settlement pattern in the Malatya Plain indicates great mobility, with an increase in small and short-lived sites (Di Nocera 2008). Slowly, a mixed agricultural and pastoral community culturally close to the Late Uruk tradition occupied Arslantepe (3000-2800 BCE, Period VI B2). A village consisting of small mudbrick houses separated by narrow perpendicular streets rose along the slopes of the mound, at the edges of a central fortified area, the function of which is still not well known (Alvaro 2010; Frangipane 2012b; Piccione et al. 2015). At the end of this period, a violent fire destroyed the village leading to the discovery of an exceptionally rich in situ assemblage sealed in the destruction layer of the mudbrick structures and outstandingly preserved. The site is possibly abandoned for a short period of time and when it is reoccupied (2800-2500 BCE, Period VI C), the political and socio-economic changes occurred in the area definitively interrupted relations with southern Mesopotamia. Independent and rather mobile groups with fragile political structure and pastoral economy inhabited the Malatya plain with single-phase settlements combining local culture with Transcaucasian/Kura-Arax elements (Conti and Persiani 1993; Frangipane 2001; Liberotti and Alvaro 2018). In such a political and demographic fragmentation, during the second half of the third millennium the nomadic pastoral component slowly integrated with the sedentary farmers permanently occupying the territory around (2500-2000 BCE, Period VID). The highly centralized Mesopotamian model had completely disappeared at this moment (Frangipane 2018). A village of few extended families of sedentary agriculturalists gradually transformed into a small town (Frangipane 2012b). The massive fortification surrounding the village suggests a certain degree of conflict within the region (Di Nocera 2008), where Arslantepe is no longer the only centre.

### 3 Methods and sampling strategy

This analytical work is part of a wider project, partially published in international peer-reviewed journals (Liberotti and Alvaro 2018; Liberotti and Quaresima 2012), which considered a greater number of mudbrick buildings and construction material samples. At first, it was undertaken as a methodological pilot study in order to investigate how the application of analytical techniques could add information to the study of earthen construction materials. Although it would be ideal to collect a minimum of three samples from each building as it is difficult to identify compositional anomalies from a single sample, for the work described in this paper only eight mudbrick samples were collected from buildings relevant to different periods of the Arslantepe architectural sequence with the purpose of examining the chronological variation. They were collected from a cleaned, fresh wall surface to avoid mineral alteration of surfaces exposed to rain, wind, or water evaporation. Mortar and mud plaster have been carefully separated from the bricks, as they often have different compositions and need an independent investigation (Love 2017). A significant factor in the sampling strategy was also the quantity and quality of the material preserved in situ as well as the presence of contextual or other published information. The structures where the samples were extracted from have also been investigated from an architectural point of view in order to better understand their original context. The studied structures are:

-A582, one of the rooms in Building 25, Period VII

-A450, the main hall of Temple B (Building 40), Period VI A

-A796, the northern section of the connecting corridor flanking Temple B, Period VI A

-A937, one of the northern rooms set within the disused fortification wall, Period VI B2

-A1886, one of the southern rooms set within the disused fortification wall, Period VI B2

-A607+, a compound of single-room dwelling units erected during Period VI C

-A529, a small room in the central area of the village, Period VI D.

The architectural analysis of the above mentioned structures has been carried out together with Dr. Corrado Alvaro (Sapienza University of Rome) and is based on topographic measurements, rectified photos and site drawings (plans, sections and elevations) of the architectural remains, in most cases exposed to a very large extent. Great attention was paid to the shape and size of the buildings, their in-house equipment, construction techniques and circulation patterns, all features that offer important clues on skills and approach to the construction and decoration of a house, on the quantity and quality of the building material used, ultimately on how people worked to create a built environment. AutoCAD software was used to elaborate handmade drawings executed on the field and generate 3D hypothetical reconstructions. Room A450 and the connected corridor A796 were also scanned through a P20 Leica Laser Scanning to easily and accurately obtain 3D information on every element of the building (Liberotti and Alvaro 2017).

The archaeometric characterization concerned two mudbrick samples for every different period spanning the Late Chalcolithic to the Early Bronze age, with the exception of Period VI C (Table 1). Although it turned out to be very significant from the architectural point of view because, as will be discussed below, a new concept of built space and new architectural models emerged during Period VI C (Liberotti and Alvaro 2018), unfortunately it was not possible to analyze any mudbrick sample from this period for this paper. For each pair of samples, one was pulverized and used for chemical and physical analysis (n. 1, 3, 5, 7); the other (n. 2, 4, 6, 8) was cut in a cube-like shape to be subjected to compression tests.

Nevertheless, only for Period VII and VI D it was possible to collect the two pairs of samples within the same room, namely A582 and A529. For Period VI A and VI B2, the two pairs of samples were extracted from one single context but not from the same room: A450, the main hall of Temple B, and its side corridor A796 for Period VI A; room A937 and room A1886, north and south of the same fortification wall, for Period VI B2. Regarding the latter, it is not certain that the two rooms were in use at the same time. Furthermore, one off-site control sample (n. 9, named as 'EXP' in the graphics below) has been collected from a mudbrick experimentally made up with soil extracted at the bottom of the mound and analyzed to compare locally occurring minerals and sediments against the culturally modified sediments found in the ancient mudbricks. Elemental analysis by X-ray fluorescence (XRF) was performed using a Spectra 2000 X-ray fluorescence spectrometer with automated sample feed to determine the chemical compositions of the samples. Helium

| | Source | Chronology | Architectural analysis | X-ray Fluorescence | Porosity | Mechanical test |
|---|---|---|---|---|---|---|
| 1 | Room A582 | Period VII | √ | √ | √ | - |
| 2 | Room A582 | Period VII | √ | - | - | √ |
| 3 | Hall A450 | Period VI A | √ | √ | √ | |
| 4 | Corridor A796 | Period VI A | √ | - | - | √ |
| 5 | Room A1886 | Period VI B2 | √ | √ | √ | - |
| 6 | Room A937 | Period VI B2 | √ | - | - | √ |
| - | Group of rooms A607+ | Period VI C | √ | - | - | - |
| 7 | Room A529 | Period VI D | √ | √ | √ | - |
| 8 | Room A529 | Period VI D | √ | - | - | √ |
| 9 | Bottom of the mound | modern | - | √ | √ | - |

Table 1. Types of analysis performed on the sampled mudbricks

pycnometry and dry flow porosimetry analyses were performed using respectively the GeoPyc 1360 instrument (Micromeritics Inc.) and the AccuPyc 1330 gas displacement pycnometer to measure density and porosity. These analytical techniques were experimentally employed at the laboratory of the University of L'Aquila by Prof. Raimondo Quaresima (Dept. of Civil, Construction-Architectural and Environmental Engineering), Dr. Fabiola Ferrante (Dept. of Industrial and Information Engineering and Economics) and Prof. Alexander Karamanov (Chemical and Physical Institute of the Bulgarian Science Academy) to determine their applicability and effectiveness on ancient earthen materials. Mechanical tests were performed at the Structures and Materials Testing Laboratory of the University of Florence by Prof. Ugo Tonietti, Dr. Luisa Rovero and Dr. Gianfranco Stipo (Dept. of Architecture) to evaluate the main mechanical parameters of the material using a hydraulic press with 50 kN loading cell and four displacement transductors type CE Cantilever (Liberotti et al. 2016). Tests were performed in displacements control in order to record the diagram load-displacement also in the post peak phase.

## 4 Architectural data

The architectural remains of Period VII consist of domestic units located in the northeast side of the site and elite residences, as well as the monumental Temple C, in the central part of the site. Both domestic units and the elite residences show round hearths. A 170 sq. m elite residence named Building 25 shows a large northeast-southwest oriented room, part of which was named A582, flanked by two smaller rooms, one of which equipped with a dome-like covered oven (Figure 2). The 120 cm thick, west wall of room A582 is made up of rows of mudbricks, whose size is irregular, especially for what concerns the length, ranging from 20 x 25 x 6 cm up to 97 x 25 x 8 cm. The considerable thickness of this wall together with the painted plaster and the columns along the walls, as well as the division of space, attest a greater building competence here than in the domestic units.

In the southwestern part of the mound, the non-domestic, northeast-southwest oriented, 165 sq. m wide Building 40 was unearthed within the so-called palatial complex of Period VI A. It is a bipartite structure used for cultic purposes (Frangipane 2018) consisting of a rectangular, 65 sq. m floor space room (A450) flanked by three smaller rooms on the long side (Figure 3). Room A450 has 120-140 cm thick walls decorated by painted mud plaster and made up by regular mudbricks

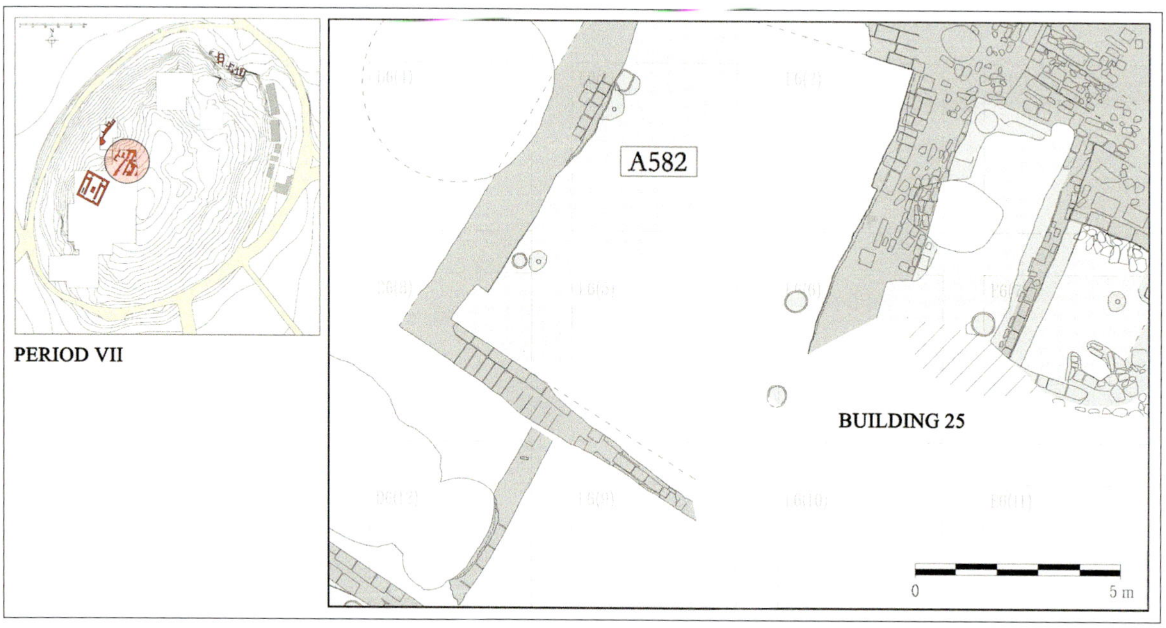


Figure 2. Key map of the Period VII settlement to the left; Building 25 with room A582 to the right.

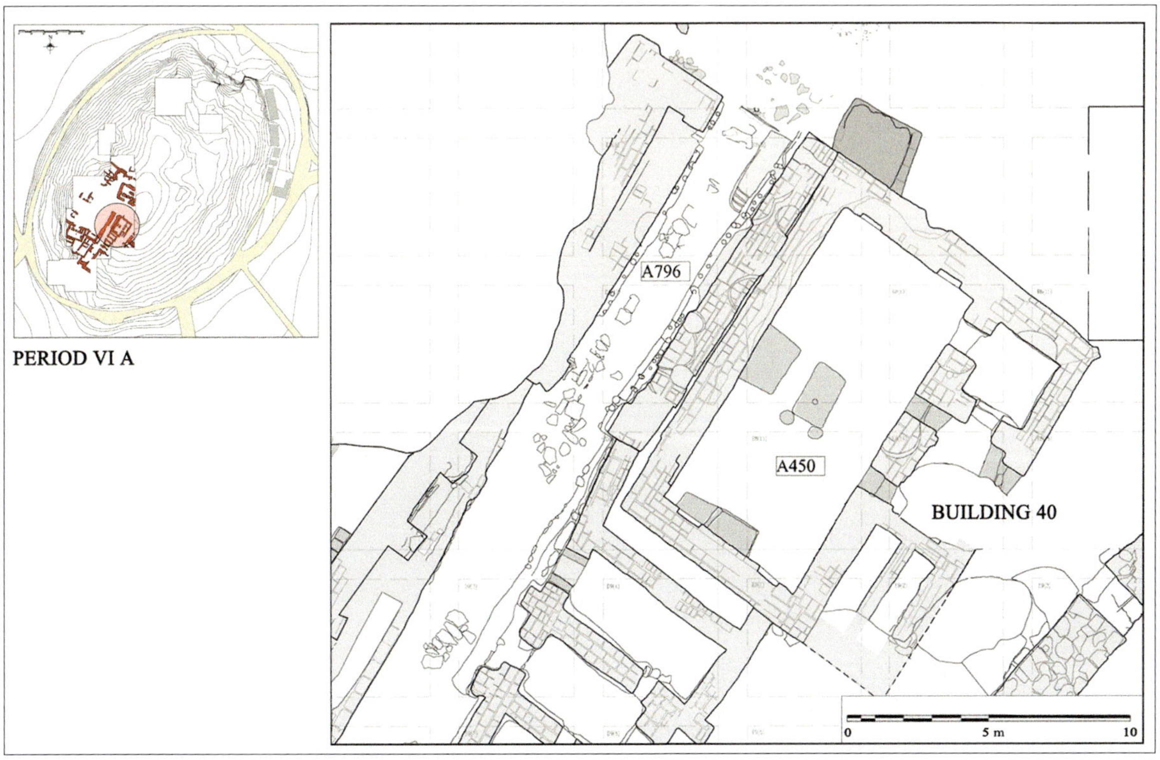


Figure 3. Key map of the Period VI A settlement to the left; Building 40 with room A450 and the connecting corridor A796 to the right.

ranging in size between 30 x 60 x 8 cm and 40 x 50 x 7 cm. The walls of the connecting corridor A796 flanking the west side of A450 are 120 cm thick and preserved up to 2 m in height. Mudbricks are very similar to those of room A450. North of A450 and A796, but to a very short distance, a group of elite residences were erected with the same space distribution, stressing an overlap between public and private buildings. For this period, there is no evidence of simple residential units or ovens so far, while hearths are the round type.

Wooden and clay huts with round hearths and a relatively large mudbrick structure characterize the first part of the subsequent Period VI B, related to the passage of Transcaucasian nomadic groups. A 5 m thick mudbrick wall on stone foundations marked the second part of this period (Period VI B2). It fortified the highest area of the mound and was furnished by three inner small rooms set within its north side. A937 is the westernmost of these rooms, measuring 13 sq. m and accessible from a door in the northeastern wall (Figure 4). At a later stage, residential units consisting of two or three rectangular small rooms connected to each other by inner courts abut to the fortification wall on its southern side. They were equipped with round hearths and domed ovens and were grouped in blocks separated by alleyways. One of the residential units was formed by A1886, a 12 sq. m room accessible from the northwestern side. Both southern and northern rooms have 35-50 cm thick pressed clay walls, whilst the mudbricks of the fortification wall, ranging in size from 48 x 30 x 8 cm to 55 x 25 x 8 cm and arranged as alternating headers and stretchers, were clearly reused as south wall of room A937 and north wall of room A1886.

In the following Period VI C, the mound is inhabited only in its central area. There is no evidence of the previous housing model nor of monumental structures built for and by the community. Quadrangular rooms with a 40 sq. m average floor space show 70 cm thick mudbrick walls on medium-sized stone foundations. All entrances are to the west and open into single rooms that do not communicate with each other. Each room shows the same type of in-house equipment, namely horseshoe-shaped hearths, irregular-shaped ovens with a small side channel for ash collection and two/three stones along the central axis that sustain poles supporting the roof. One of these

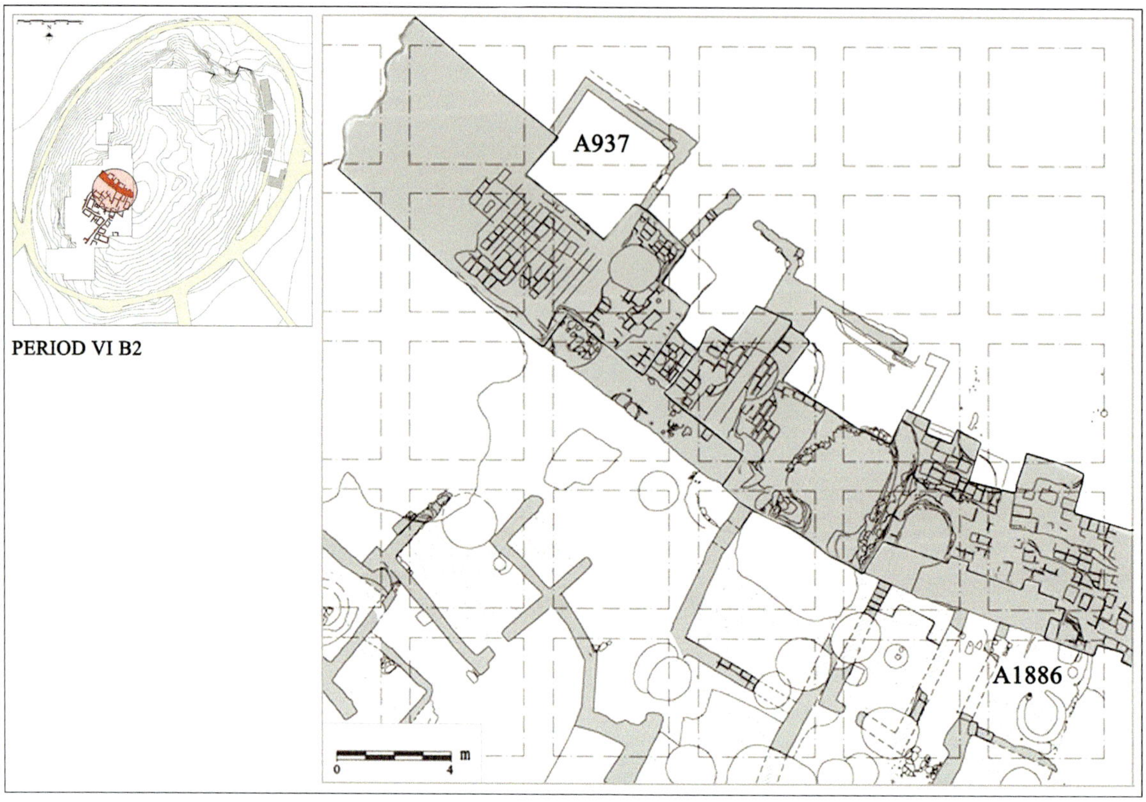


Figure 4. Key map of the Period VI B2 settlement to the left; fortification wall and rooms A937 and A1886 to the right.

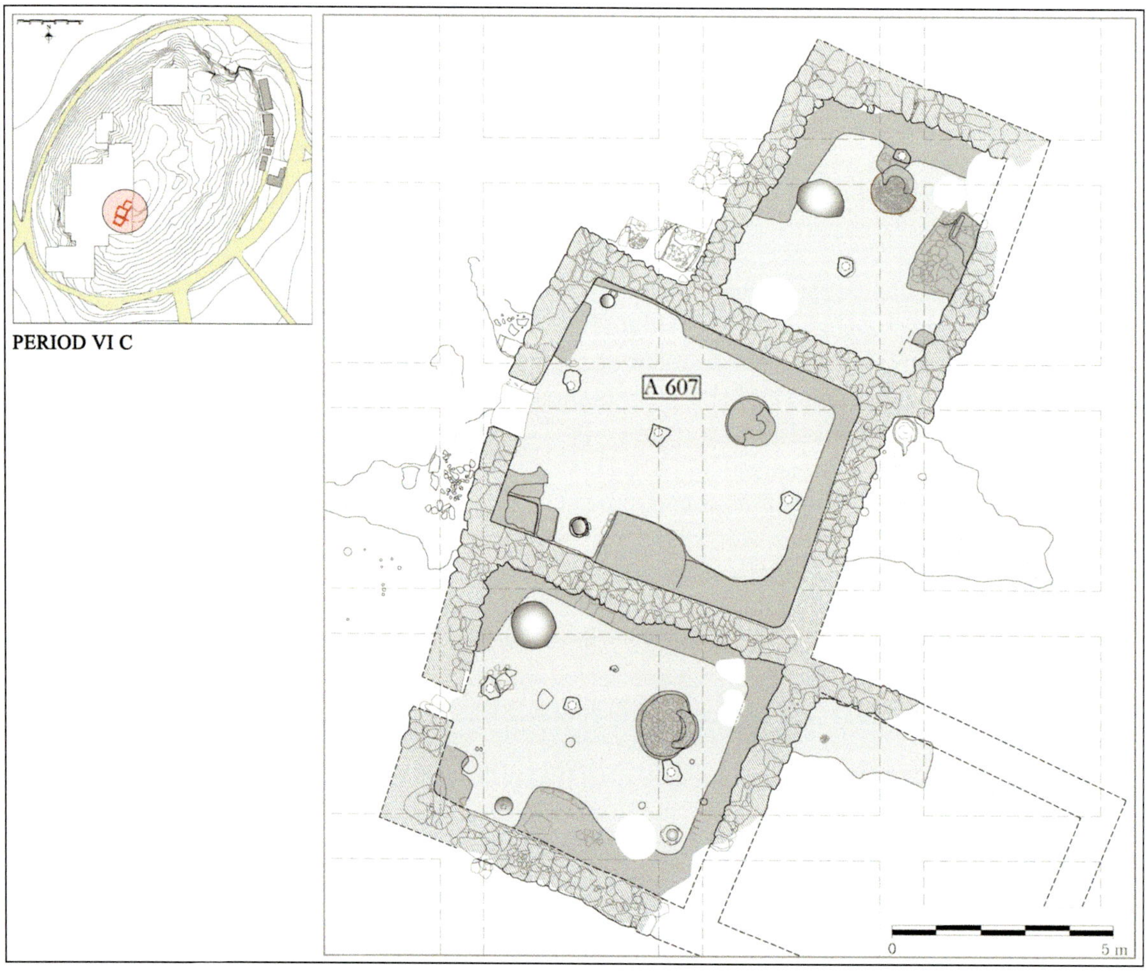


Figure 5. Key map of Period VI C to the left; the A607+ compound to the right.

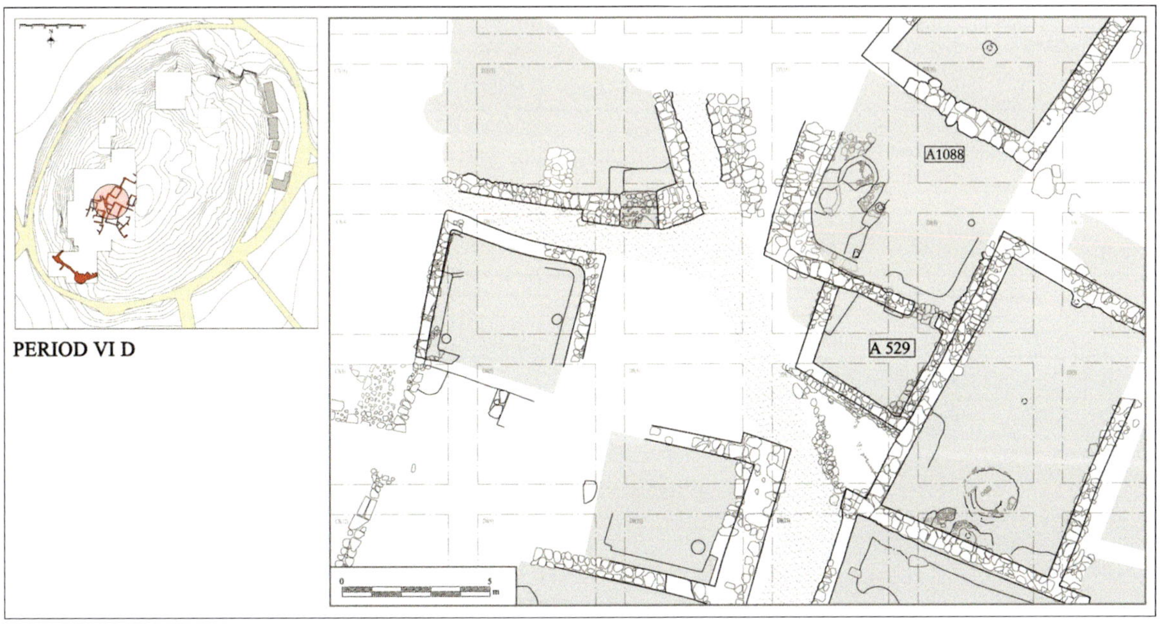


Figure 6. Key map of one of the period VI D settlement to the left; rooms A529 and A1088 to the right.

rooms, A607, shows a 100 x 80 cm single stone threshold and a quadrangular mudbrick platform abutting a bench that runs along all walls (only partially on the west wall) (Figure 5). Both platform and bench consist of 45/60 x 30 cm homogeneous mudbricks laid in a stretcher and header course combination, sometimes on the long narrow side. They enable to understand how not-preserved external walls had to be.

In the following Period VI D, the above-mentioned compound expands. New 40 sq. m rooms and open court are built. Access to the rooms is now from the east side. Stone foundation, wall thickness, benches, ovens and hearths are identical to those of Period VI C. No monumental structures have been unearthed so far. The inhabited area further expands over time. An imposing fortification wall on stone foundation, partially unearthed in the southern area, surrounds the village. Residential units consist of two or three rooms with a floor space of 40/50 sq. m. Terraces, water drainage systems and alleys separating groups of residential units are built. A529 is an 11 sq. m room of one residential unit within the village. It is connected through a door to room A1088, which is possibly relevant to the same residential unit. Mudbricks are large and regular in size, 60/65 x 45 cm (Figure 6).

## 5 Archaeometric data

This section presents the results of chemical, physical and mechanical analyses (Table 2). The X-ray fluorescence analysis (Figure 7) show that the elemental composition of the eight mudbrick samples extracted from ancient buildings as well as of the sample extracted from the experimental mudbrick is similar, with some variation in the percentage of element content.

The variation may explain the dissimilarity in colour, which range from dark to light grey and brown to red especially in Period VII and VI A mudbricks, rather than a discrepancy in the mudbrick performance. Colour variation and its origin from post-depositional processes (raw material) rather than from production techniques (tempering agents added during manufacture) has been discussed elsewhere (Liberotti 2012; Liberotti and Quaresima 2010).

Although the elemental profiles is similar in all samples, a critical dissimilarity was observed in sample n. 1, where Chloride and Sulphur are present in much higher concentrations. The Chloride value is more than four times higher than that of the other samples. The Sulphur value is nearly

| | Al% | Fe% | Si% | Ca% | Mg% | K% | Cl% | S% | Porosity% | MPa |
|---|---|---|---|---|---|---|---|---|---|---|
| 1 | 6,19 | 5,90 | 29,63 | 40,06 | 7,74 | 6,46 | 2,18 | 0,75 | 29,03 | - |
| 2 | - | - | - | - | - | - | - | - | - | 0,18 |
| 3 | 6,11 | 6,95 | 31,79 | 42,77 | 6,57 | 3,71 | 0,57 | 0,13 | 47,98 | - |
| 4 | - | - | - | - | - | - | - | - | - | 1,64 |
| 5 | 5,14 | 6,40 | 29,34 | 48,89 | 5,11 | 3,13 | 0,37 | 0,15 | 44,72 | - |
| 6 | - | - | - | - | - | - | - | - | - | 0,31 |
| 7 | 4,97 | 6,67 | 28,52 | 47,35 | 6,20 | 6,00 | 0,72 | 0,19 | 45,30 | - |
| 8 | - | - | - | - | - | - | - | - | - | 0,70 |
| 9 | 5,50 | 6,31 | 31,87 | 44,87 | 5,92 | 3,31 | 0,65 | 0,14 | 44,45 | - |

Table 2. Chemical, physical and mechanical raw data from the tested samples

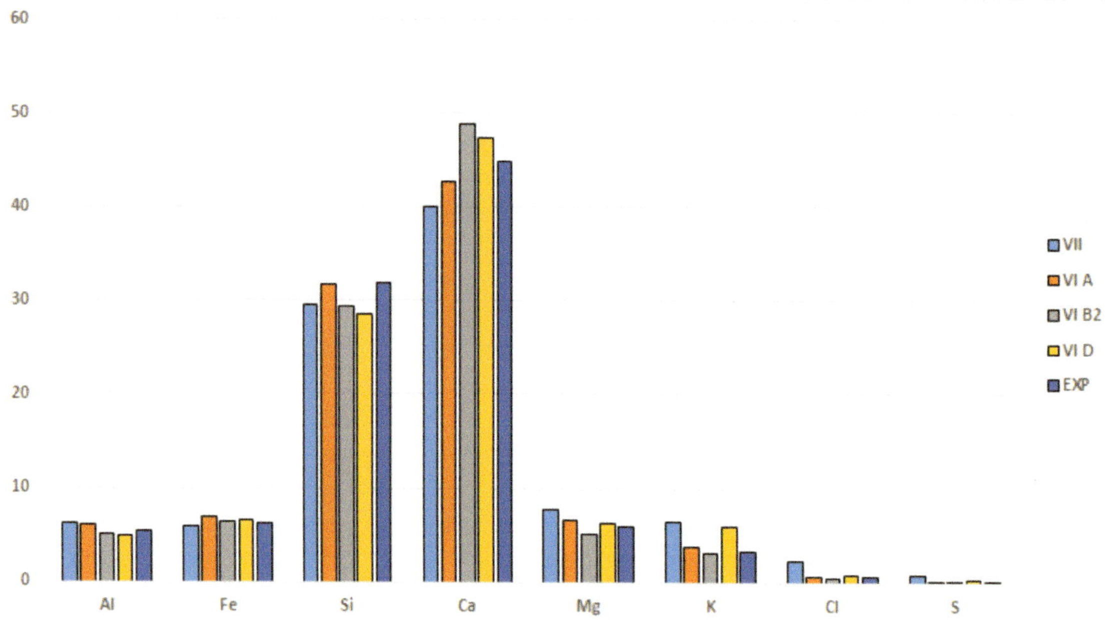


Figure 7. Results of the X-ray Fluorescence analysis.

five times higher than that of the other samples (Figure 8). It should be observed here that seven more mudbrick samples from the same room A582 have been analysed by X-ray fluorescence for other scientific purposes (Liberotti and Quaresima 2012: 449) and they all show the same much higher values of Chloride and Sulphur.

Another significant dissimilarity concerns the porosity values, and once again the real difference is between the Period VII sample and all the others, the former having low porosity with respect to the latter (Figure 9). Even in this case, the other seven samples collected from the same room A582 show less than 40% porosity whereas in the samples from other periods, as well as in the one extracted from the experimental mudbrick, porosity is always more than 40%. Finally, despite the good deformation capacity, the low value of compressive strength in the Period VII sample shows poor resistance to compression in comparison with the samples of the other three periods. In this regard, it may be appropriate to state that the much greater load bearing capacity, strength and stiffness showed by the Period VI A sample n. 4 could be related to the uncontrolled exposure to fire.

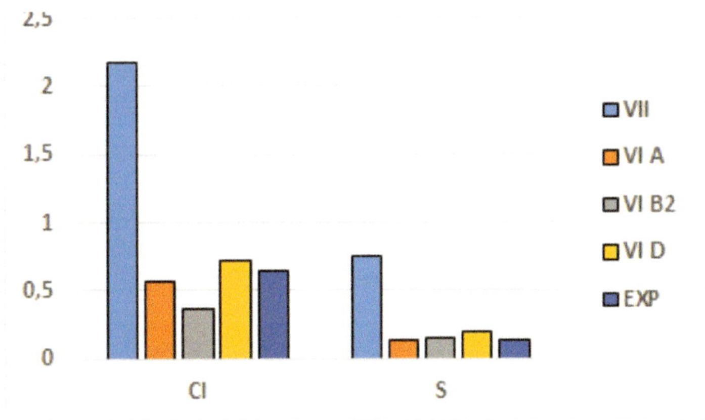


Figure 8. Comparison of Chloride and Sulphur values.

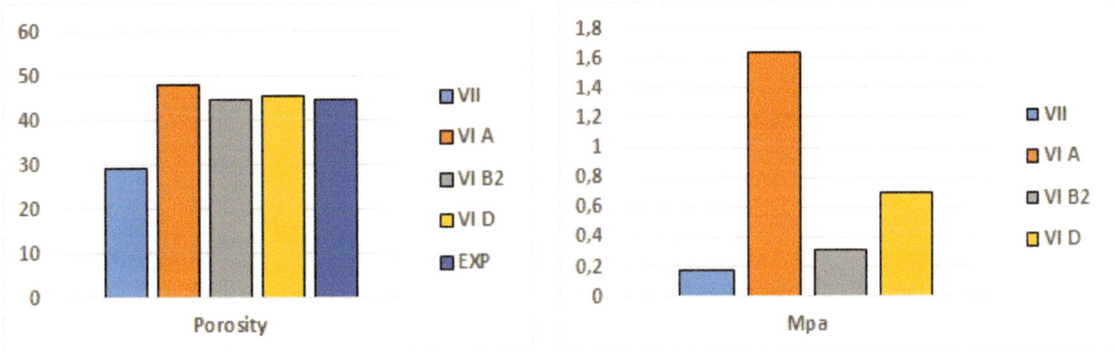


Figure 9. Comparison of porosity values to the left, of compressive strength values to the right.

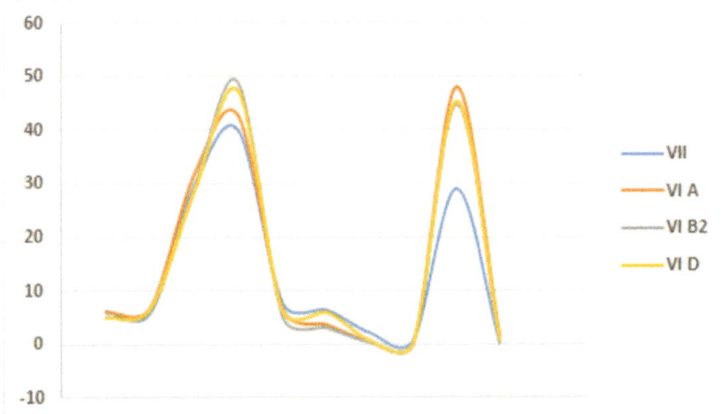


Figure 10. Plot of all samples and analyses.

Taking into account the results of all analyses, despite a generic homogeneity, samples of Period VII tend to differentiate slightly from the others, especially in the content of certain minor chemical elements, in the porosity value and in the resistance to compressive strength. Although in the graph below the light blue line, which corresponds to the Period VII sample, follows the general trend, it moderately declines at the Calcium and Porosity peaks and rises a little at the Magnesium, Chloride and Sulphur peaks (Figure 10). The other lines maintain the same level.

## 6 Discussion

The long-lasting extensive archaeological excavation at Arslantepe provided an enormous amount of information about the architectural evidence of several different settlements spread over millennia. This rich data set acts as a pivot in the interpretation of each single building and establishes a baseline for intra-site comparison, from which substantial conclusions can be made. Despite the sometime-radical changes occurred in each building phase and the different dimensional and building parameters, elements of continuity can be followed in the architectural layout of domestic architecture of Period VII, Period VI A and Period VI B. The large thickness of the elite residence walls contrasts with the small one in the residential units. The texture of the walls is not always homogeneous, mainly because of the functional difference of the rooms. The size of the mudbricks and their arrangement within the walls are regular in the elitist and monumental building whereas it is less regular in the domestic units, even if there is some deviation (see the northwestern wall in room A582, Period VII). The use of the inner space and the settlement organisation, as well as the cooking installations (Balossi 2015), pottery (Frangipane 2012b) and metallurgical production (Palmieri et al. 1999), changed drastically when the site was abandoned after Period VI B2, marking the end of the above-mentioned continuity. During Period VI C, new groups with a rather different constructive tradition settled down in Arslantepe starting a long sequence of settlements, which last across the whole Period VI D. The wall thickness does not change from one room to another as well as the texture of the mudbricks and their size. Although the number of samples selected for a diachronic comparison is not sufficient to make any firm statement, the archaeometric analysis carried out to date seems to counteract the scenario described above. As a matter of fact, mudbricks of Period VII, and not of Period VI D, show a slight but clear differentiation in the presence of some minor element components, in their porosimetric profile and in their resistance to compressive strength. First, although the higher content of Chloride and Sulphur does not segregate mudbricks into any meaningful patterns reflecting natural variation of the source materials rather than a discrimination due to the manufacturing process, it may suggest the use of different local sources during Period VII, in an attempt to find the best fit for the planned structure. Secondly, given that porosity is closely linked to the content of temper, in this case straw, added to the mixture of the mudbricks to reduce shrinkage, facilitate workability and lighten their weight, its low value within the same Period VII samples speaks of a different approach, maybe not yet standardized, of mudbrick manufacturing. Lastly, the results of the mechanical test show that the production process of period VII mudbricks is not as accurate as that of Periods VI B2 and VI D, and definitively less accurate than Period VI A mudbricks. Perhaps the archaeometric analysis of the Period VI C mudbricks, whose archaeological evidence can be excavated again in the future, will assess this issue.

## 7 Conclusion

At this stage, the phase-by-phase analysis of the spatial and temporal distribution of building materials creates a comprehensive interpretation of mudbrick architecture at Arslantepe. This pilot intra-site study showed that combining analytical techniques with architectural investigation can add a great value to the understanding of earthen construction materials. The *chaîne opératoire* of mudbricks is not discernible in the minerals or geochemistry but their combinations, dimensions and fabric, as well as social technology, resource selection and skill of the mudbrick makers must be considered to provide insights into specialized production, standardization, and the social mechanisms driving the practice of construction of ancient societies.

## Bibliography

Aksoy, E., Türkmen, I. and Turan, M. (2005) 'Tectonics and sedimentation in convergent margin basins: an example from the Tertiary Elazığ basin, Eastern Turkey', *Journal of Asian Earth Sciences*, 25(3): 459-472.

Alvaro, C. (2010) 'Architecture and the organisation of space' in Frangipane, M. (ed.) *Economic Centralisation in Formative States. The Archaeological Reconstruction of the Economic System in 4th Millennium Arslantepe. Studi di Preistoria Orientale 3.* Roma: Sapienza Università di Roma: 45-72.

Alvaro, C. (2012) 'The topography and architecture at Arslantepe during the second and first millennia BC: reconsidering more than 100 years of researches', *Origini*, XXXIV: 345-360.

Alvaro, C. and Palumbi, G. (2014) 'New 'pastoral landscapes' of the early 3rd Millennium BC in the Malatya Plain. The case of Gelinciktepe' in Bieliński, P., Gawlikowski, M., Koliński, R., Ławecka, D.,

Sołtysiak, A. and Wygnańska, Z. (eds) *Proceedings of the 8th International Congress on the Archaeology of the Ancient Near East.* Warsaw: University of Warsaw: 607-617.

Arıkan, B., Balossi Restelli, F. and Masi, A. (2016) 'Comparative modeling of Bronze Age land use in the Malatya Plain (Turkey)', *Quaternary Science Reviews*, 136: 122-133.

Arnold, D.E. (2000) 'Does the standardization of ceramic pastes really mean standardization?', *Journal of Archaeological Method and Theory*, 7(4): 333-375.

Balossi Restelli, F. (2015) 'Hearth and home. Interpreting fire installations at Arslantepe, Eastern Turkey, from the fourth to the beginning of the second millennium BCE', *Paléorient*, 41(1): 127-151.

Baneschi, I., Masi, A., Sadori, L., and Zanchetta, G. (2012) 'The use of stable carbon isotopes in palaeoenvironmental studies in archaeology: the example of Arslantepe (Malatya, Eastern Anatolia) from 5300 to 3950 years BP', *Rendiconti Online della Società Geologica Italiana* 18: 8-11.

Bozkaya, Ö., Yalçin, H., Başibüyük, Z., Özfırat, O. and Yılmaz, H. (2007) 'Origin and evolution of the southeast Anatolian metamorphic complex (Turkey)', *Geologica Carpathica*, 58(3): 197-210.

Coffman, R., Agnew, N., Austin, G. and Doehne, E. (1990) 'Adobe Mineralogy: Characterization of Adobes from around the World' in Agnew, N., Taylor, M. and Alva Balderrama, A. (eds) *Proceedings of the 6th International Conference on the Conservation of Earthen Architecture, Adobe 90 Preprints.* Los Angeles: The Getty Conservation Institute: 424-429.

Conti, A.M. and Persiani, C. (1993) 'When worlds collide. Cultural developments in Eastern Anatolia in the Early Bronze Age' in Frangipane, M., Hauptman, H., Liverani, M., Matthiae, P., and Mellink, M. (eds), *Between the Rivers and over the Mountains. Archaeologica Anatolica et Mesopotamica Alba Palmieri Dedicata.* Roma: Università di Roma 'La Sapienza': 361-413.

Di Nocera, G.M. (2008) 'Settlements, population and landscape on the Upper Euphrates between V and II millennium BC. Results of the Archaeological Survey Project 2003-2005 in the Malatya Plain' in Córdoba, J.M., Molist, M., Pérez, M.C., Rubio, I. and Martínez, S. (eds) *Proceedings of the 5th International Congress on the Archaeology of the Ancient Near East.* Madrid: Universidad Autónoma de Madrid ediciones: 633-645.

Dreibrodt, S., Lubos, C., Lomax, J., Sipos, G., Schroedter, T. and Nelle, O. (2014) 'Holocene landscape dynamics at the tell Arslantepe, Malatya, Turkey. Soil erosion, buried soils and settlement layers, slope and river activity in a middle Euphrates catchment', *Holocene*, 24(10): 1351-1368.

Emery, V.L. (2011) 'Mudbrick Architecture' in Wendrich, W. (ed.) *Encyclopedia of Egyptology.* Los Angeles: UCLA

Forget, M.C.L., Regev, L., Friesem, D.E. and Shahack-Gross, R. (2015) 'Physical and mineralogical properties of experimentally heated chaff-tempered mud bricks: Implications for reconstruction of environmental factors influencing the appearance of mud bricks in archaeological conflagration events', *Journal of Archaeological Science: Reports*, 2: 80-93.

Frangipane, M. (1997) 'A 4th millennium temple/palace complex at Arslantepe-Malatya. North–south relations and the formation of early state societies in the northern regions of Greater Mesopotamia', *Paléorient*, 23(1): 45-73.

Frangipane, M. (2001) 'The transition between two opposing forms of power at Arslantepe (Malatya) at the beginning of the 3rd Millennium', *Tüba-Ar*, IV: 1-24.

Frangipane, M. (2009) 'Non-urban hierarchical patterns of territorial and political organisation in Northern regions of Greater Mesopotamia: Tepe Gawra and Arslantepe' in Butterlin, P., (ed.) *À propos de Tepe Gawra, le monde proto-urbain de Mésopotamie. Subartu XXIII.* Turnhout: Brepols: 135-148.

Frangipane, M. (2010) (ed.) *Economic Centralisation in Formative States: The Archaeological Reconstruction of the Economic System in 4th Millennium Arslantepe. Studi di Preistoria Orientale 3.* Roma: Università 'La Sapienza' di Roma.

Frangipane, M. (2012a) 'Fourth Millennium Arslantepe: The Development of a Centralised Society without urbanization', *Origini*, XXXIV: 19-40.

Frangipane, M. (2012b) 'The Collapse of the 4th Millennium Centralised System at Arslantepe and the Far-Reaching changes in the 3rd Millennium Societies', *Origini*, XXXIV: 237-260.

Frangipane, M. (2018) 'Different Trajectories in State Formation in Greater Mesopotamia: A View from Arslantepe (Turkey)', *Journal of Archaeological Research*, 26(1): 3-63.

Frangipane, M. and Palmieri, A. (1983) 'Perspectives on Protourbanization in Eastern Anatolia: Arslantepe (Malatya). An Interim Report on 1975–1983 Campaigns. A Protourban Centre of the Late Uruk Period', *Origini*, XII (2): 287-454.

French, C.A.I. (1984) 'Sediments Analysis of Mud Brick and Natural Features at El-Amarna' in Kemp, B. (ed.) *Amarna Reports I.* London: Egypt Exploration Society.

Homsher, R.S. (2012) 'Mud bricks and the process of construction in the Middle Bronze Age southern Levant', *Bulletin of the American Schools of Oriental Research*, 368: 1-27.

Kaya, Z. and Raynal, D.J. (2001) 'Biodiversity and conservation of Turkish forests', *Biological Conservation*, 97(2): 131–141.

Kemp, B. (2000) 'Soil (Including Mudbrick Architecture)' in Nicholson, P. T. and Shaw, I. (eds) *Ancient Egyptian Materials and Technology.* Cambridge: Cambridge University Press: 78-103.

Liberotti, G. (2012) 'Architectural and Archaeometric Analysis of Buldings and Building Materials from Late Chalcolithic Arslantepe (Malatya, Turkey)' in Matthews, R. and Curtis, J. (eds) *Proceedings of the 7th International Congress on the Archaeology of the Ancient Near East. Vol. 3. The British Museum and University College of London.* Wiesbaden: Harrassowitz Verlag: 579-592.

Liberotti, G. and Alvaro, C. (2017) 'Using laser scanner technology to analyse mudbrick architecture in the ancient Near East. The palatial complex of Arslantepe (Malatya, Turkey)', *Archeologia e Calcolatori*, 28(2), pp 447-455.

Liberotti, G. and Alvaro, C. (2018) 'Arslantepe-Malatya: domestic architecture and building technology during the Early Bronze Age II-III', *Scienze dell'Antichità*, 24(1) pp. 19-37.

Liberotti, G. and Quaresima, R. (2010) 'Building materials in the 4th and early 3rd millennium monumental architecture at Arslantepe: mud bricks and plaster' in Frangipane, M. (ed.) *Economic Centralisation in Formative States. The Archaeological Reconstruction of the Economic System in 4th Millennium Arslantepe. Studi di Preistoria Orientale 3.* Roma: Sapienza Università di Roma: 73-80.

Liberotti, G. and Quaresima, R. (2012) 'Building material and construction techniques at Arslantepe: results of an interdisciplinary study', *Origini*, XXXIV: 447-463.

Liberotti, G., Rovero, L., Stipo, G., and Tonietti, U. (2016) 'Mechanical investigation on adobe samples belonging to the archeological site of Arslantepe (Malatya, Turkey)', *Journal of Mechanical Engineering & Sciences*, 7(10): 3656-3666.

Love, S. (2013) 'The Performance of Building and Technological Choice Made Visible in Mudbrick Architecture', *Cambridge Archaeological Journal*, 23(2), pp 263-282.

Love, S. (2017) 'Field Methods for the Analysis of Mud Brick Architecture', *Journal of Field Archaeology*, 42(4): 351-363.

Marcolongo, B. and Palmieri, A.M. (1983) 'Environment, water supply and cultural development at Arslantepe (Malatya, Turkey)', *Origini*, 12(2): 619-628.

Masi, A., Sadori, L., Balossi Restelli, F., Baneschi, I. and Zanchetta, G. (2014) 'Stable carbon isotope analysis as crop management indicator at Arslantepe (Malatya, Turkey) during the Late Chalcolithic and Early Bronze Age', *Vegetation History and Archaeobotany*, 23(6): 751–760.

Nodarou, E., Frederick, C. and Hein, A. (2008) 'Another (Mud) Brick in the Wall: Scientific Analysis of Bronze Age Earthen Construction Materials from East Crete', *Journal of Archaeological Science*, 35(11): 2997-3015.

Önal, A. (2008) 'Mineral chemistry, crystallization condition and magma mixing–mingling at Orduzu Volcano (Malatya), Eastern Anatolia, Turkey', *Geological Journal*, 43(1): 95-116.

Önal, A., Boztuğ, D., Arslan, M., Spell, L.T. and Kürüm, S. (2008) 'Petrology and 40Ar/39Ar age of the bimodal Orduzu volcanic (Malatya) from the western end of the Eastern Anatolian Neogene volcanism, Turkey', *Turkish Journal of Earth Sciences*, 17: 85-109.

Palmieri, A.M. (1973) 'Scavi nell'area sud-occidentale di Arslantepe. Ritrovamento di una struttura templare dell'Antica Età del Bronzo', *Origini*, 7: 55-228.

Palmieri, A.M. (1978) 'Studio sedimentologico dell'area nord-occidentale di Arslantepe (Malatya, Turchia)', *Quaderni della Ricerca Scientifica*, 100: 353-364.

Palmieri, A.M., Frangipane, M., Hauptmann, A. and Hess, K. (1999) 'Early Metallurgy at Arslantepe during the Late Chalcolithic and the Early Bronze Age IA-IB Periods' in Hauptmann, A., Pernicka, E., Rehren, T. and Yalçin, Ü. (eds) *The Beginnings of Metallurgy. Proceedings of the International Conference.* Bochum: Deutsches Bergbau-Museum: 141-148.

Palumbi, G. (2012) 'Bridging the Frontiers. Pastoral Groups in the Upper Euphrates Region in the Early Third Millennium BCE', *Origini*, XXXIV: 261-278.

Piccione, P., Alvaro, C., Bartosiewicz, L., Lemorini, C., Masi, A. and Sadori, L. (2015) 'Distribution of artifacts and ecofacts in an Early Bronze Age house in Eastern Anatolia: Space use and household economy at Arslantepe VI B2 (2900-2750 BCE)', *Journal of Archaeological Science: Reports*, 4: 8-22.

Pritchard, J.B. (1985) *Tell es-Sa'idiyeh. Excavations on the Tell 1964-66.* Philadelphia: The University Museum, University of Pennsylvania.

Puglisi, S.M. and Meriggi, P. (1964) *Malatya I. Rapporto preliminare delle campagne 1961 e 1962.* Roma: Centro per le Antichità e la Storia dell'Arte del Vicino Oriente.

Robertson, A.H.F., Ustaömer, T., Parlak, O., Ünlügenç, U.C., Tasli, K. and İnan, N. (2006) 'The Berit transect of the Tauride Thrust Belt, S. Turkey: Late Cretaceous - Early Cenozoic accretionary/collisional processes related to closure of the southern Neotethys', *Journal of Asian Earth Sciences*, 27(1): 108-145.

Rosen, A. (1986) *Cities of Clay: The Geoarchaeology of Tells.* Chicago: University of Chicago Press.

Sadori, L. and Masi, A. (2012) 'Archaeobotanical research at Arslantepe: traditional approach and new challenges', *Origini*, XXXIV: 433-446.

Sauvage, M. (2011) 'L'architecture de brique crue en Mésopotamie' in Chazelles, C.A., Guillaud, H. and Klein, A. (eds) *Les cultures constructives de la brique crue. Troisièmes échanges transdisciplinaires sur les constructions en terre crue.* Montpellier: Éditions de l'Espérou: 89-100.

Schneider Equini, S. (1970) *Malatya II. Rapporto preliminare delle campagne 1963-1968. Il livello romano bizantino e le testimonianze islamiche.* Roma: Centro per le Antichità e la Storia dell'Arte del Vicino Oriente.

Stevanovic, M. (2003) 'Building an experimental house at Çatal Höyük' in Bellintani: and Moser, L. (eds) *Proceedings of Archeologie sperimentali, metodologie ed esperienze fra verifica, riproduzione, comunicazione e simulazione.* Trento: Comano Terme o Fiavè: 253-271.

Yazgan, E. and Mason, R. (1988) 'Orbicular gabbro near Baskil, southeastern Turkey', *Mineralogical Magazine*, 52(365): 161-173.

Yiğitbaş, E. and Yılmaz, Y. (1996) 'New evidence and solution to the Maden Complex controversy of the southeast Anatolian orogenic belt (Turkey)', *Geologische Rundschau*, 85(2): 250-263.

# Chapter 5

# Minoan mudbricks: earth and fire in Bronze Age Crete

Marta Lorenzon
The University of Helsinki (Centre of Excellence Ancient Near Eastern Empires)
& Wiener Laboratory, American School of Classical Studies at Athens
marta.lorenzon@helsinki.fi

**Abstract**

This contribution focuses on the SEM and FT-IR analyses of Bronze Age earthen architecture in Crete in order to investigate its main characteristics and to advance hypotheses regarding the lack of long-term preservation. Specifically, this study employed geoarchaeological techniques to assess the nature and the effects of the depositional and post-depositional processes that affect Minoan mudbricks. Its main goal is to highlight ways in which geoarchaeology not only informs archaeologists on raw source material procurement, technological developments, and craft specialization, but it also provides a powerful tool to understand the preservation history of earthen constructions and to examine the main causes of decay.

**Keywords:** Microstructure analysis, Preservation, Geoarchaeology, FT-IR, SEM, Bronze Age Aegean

**Résumé**

Cette contribution se concentre sur les analyses SEM et FT-IR de l'architecture en terre de l'âge du bronze en Crète afin d'en explorer les caractéristiques principales et de faire des hypothèses sur le manque de préservation à long terme. Plus précisément, cette étude a utilisé des techniques géoarchéologiques pour évaluer la nature et les effets des processus de déposition et de post-déposition qui affectent les briques de terre crue Minoennes. L'objectif principal est de mettre en évidence les moyens par lesquels la géoarchéologie informe les archéologues non seulement sur l'approvisionnement en matières premières brutes, les développements technologiques et la spécialisation artisanale, mais fournit également un outil puissant pour comprendre l'histoire de la préservation des constructions en terre et pour examiner les principales causes de dégradation.

**Mots-clés:** analyse de microstructure, préservation, géoarchéologie, IRTF, MEB, âge dU bronze égéen

## 1 Introduction

The preservation of earthen architecture in archaeological contexts depends on multiple factors, such as climate, site depositional process, and treatment of the architectural material once excavated. The relevance of earthen building materials both as ecofacts and artefacts has been increasingly recognized, as mudbrick studies reveal them to be a rich source of information for archaeological investigation, providing data on the natural environment, labour organization and building practices (Friesem et al. 2014; Lorenzon and Iacovou 2019; Love 2012 2017). Until recently though, Bronze Age mudbricks from Crete associated with Minoan architecture were not part of the investigated archaeological materials, mostly due to poor preservation and diagnostic problems (Guest-Papamanoli 1978; Jerome 1993; Lorenzon 2017; Nodarou et al. 2008). Recent studies on these Bronze Age mudbricks have used both macroscopic and microscopic analysis to shed light on issues of raw source procurement, technological development, and craft specialization in relation to public and domestic architecture (Devolder 2009; Devolder and Lorenzon 2019; Guest-Papamanoli 1978; Lorenzon 2017; Nodarou et al. 2008). The principal challenge facing the analysis of Minoan earthen architecture is two-fold: A) the identification of earthen building materials during excavation, which is often problematic in the Aegean environmental conditions, and B) the nature of the preserved material, which can increase the difficulties in extracting information through traditional geoarchaeological methods (Devolder and Lorenzon 2019; Lorenzon 2017; Nodarou et al. 2008).

First, Aegean environmental conditions make unfired earthen building materials extremely difficult to preserve *in situ* (i.e., earthen materials' decay is often caused by erosion, temperature variations, and capillary moisture) and often the collapsed material is not recognized during excavation as the rubble is highly affected by post-depositional processes (Karkanas and Van der Moortel 2014; Kontogiorgos 2010; Nodarou et al. 2008; Goodman-Elgar 2008; McIntosh 1974). Second, most of the Minoan material still preserved seems to have undergone some sort of fire treatment, impacting the clay's crystalline structure and the properties of the structural material, and thus affecting the type of information that could be collected from traditional earthen architecture analytical studies (i.e., hydrometer analysis, vegetal temper identification) (Forget et al. 2015; Lorenzon 2017; Marocchi et al. 2010).

This contribution investigates the results of two specific analytical techniques applied to Minoan earthen building materials, scanning electron microscopy (SEM) and Fourier transform infrared spectrometry (FT-IR), indicating how these techniques can shed light on the material's key structural characteristics. The 120 samples analyzed were collected from three Bronze Age sites in Crete, Gournia, Malia and Monastiraki, in order to assess their performance, microstructure properties and preservation history.

## 2 Archaeological background, methods and results

The presence of earthen building materials has been attested in the most relevant Bronze Age sites in Crete, such as Gournia, Malia, Knossos, Monastiraki, Palaikastro, Kato Zakro and Chania (Boyd-Hawes et al. 1908; Bradfer-Burdet and Pomadère 2011; Devolder 2009; Devolder and Lorenzon 2019; Jerome 1993; Shaw 2009). Bronze Age Crete is associated with the emergence and development of the Minoan civilization, thus the traditional chronological division of Early, Middle and Late Bronze Age (EBA, MBA and LBA) is often mimicked in the early archaeological publications as Early, Middle, and Late Minoan periods or in a division grounded in different architectural palace phases (i.e., Prepalatial, Protopalatial, Neopalatial and Postpalatial). More recently absolute dating based on re-defined high and/or low chronology has been discussed and has been increasingly adopted in the archaeological literature (Bietak 2014; Manning et al. 2014; Warren 2010). This contribution adopts – whenever possible – the Bronze Age terminology and focuses on the mudbrick samples collected from the sites of Monastiraki, characterized by MBA occupation and of Gournia and Malia, which both presented EBA, MBA and LBA occupation (Figure 1).

These three sites were selected because they shared two important characteristics: they presented long-term occupational patterns during the Bronze Age and showed extensive use of earthen constructions within their urban environment. The 120 samples collected with the permission of the Ministry of Culture and respective Eforias, were put through a series of analytical processes to answer comprehensive research questions related to raw source procurement, labour organization and craft specialization (Lorenzon 2017; Devolder and Lorenzon 2019). This paper as an original publication focuses on presenting the results of two main techniques in relation to material performance and preservation:

- Fourier transform infrared spectrometry (FT-IR) was employed to detect effects of temperature changes on clay minerals (for methodological approaches and results see Berna et al. 2007; Friesem et al. 2014; Forget et al. 2015). The instrument used was a Thermo Scientific Model Nicolet iS5 with a spectral range 7800-500 cm-1. The samples were ground with KBr powder and then pressed into disks for the analysis (for the detailed methodology see Weiner 2010). The results were then compared with reference samples.
- Scanning electron microscopy (SEM) was used to characterize the sample microstructure and to assess differences in densities and porosity (Jerome 1993; Singh et al. 2016, Zhang et al. 2018). SEM analysis was performed with a JEOL JMS-IT300LV instrument at the Wiener Laboratory for Archaeological Science in Athens.

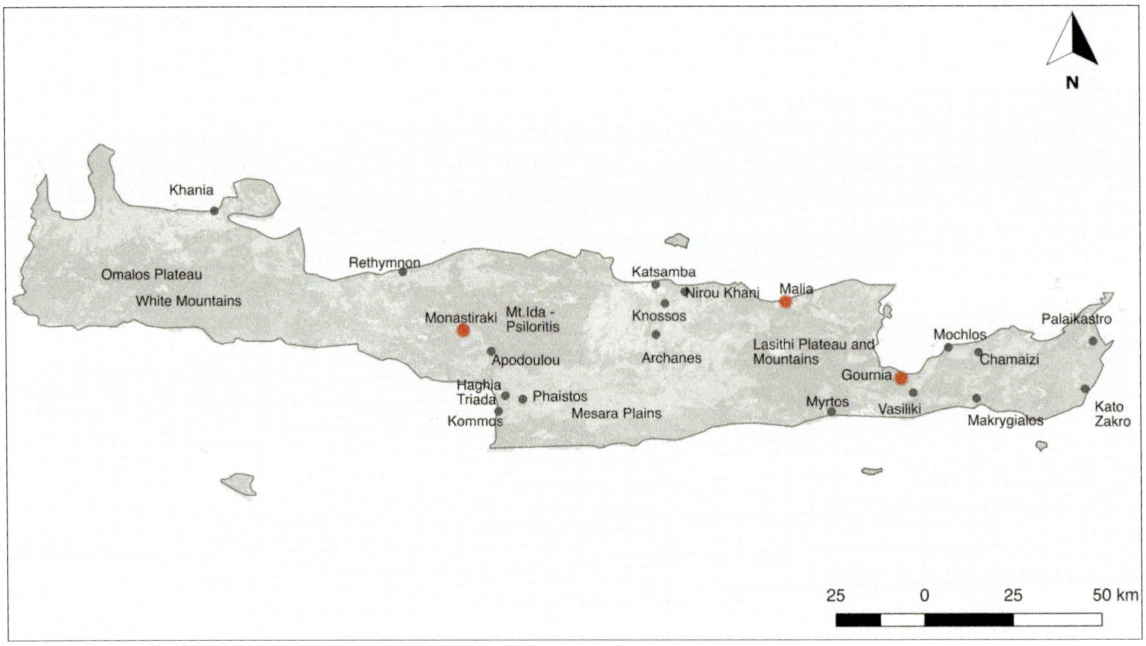


Figure 1. Map of Crete with the three sites' location (Data Source: Natural Earth)

FT-IR spectra indicate that all the Bronze Age samples have experienced a heat treatment, although they have been subjected to widely different temperatures. Specifically, the absence of structural water (peak at 3600 cm-1) and various shifts of the main clay peaks (clay indicative absorptions at 1035, 913, 798, 778 cm-1) are directly related to the intense heat developed during conflagration, but the temperature of the conflagrations was very different in all the samples analyzed (Figure 2). FT-IR spectra also denote the heterogeneous nature of Minoan mudbricks inclusions, which comprise calcite, iron oxides and calcium hydroxides. These inclusions are visible in multiple samples, as clear peaks of calcite at 1430, 875 cm-1; iron oxide at 694 cm-1 peak; and calcium hydroxide at 3621 cm-1 peak. Iron oxides are part of the natural geology of the Terra Rossa soil, which is often used in Minoan earthen construction and it is also well discernible at the macroscopic level due to its revealing red colour (Devolder 2009; Jerome 1993; Lorenzon 2017). Additionally, iron oxides also contribute to Terra Rossa's main mechanical characteristics, such as the tight microstructure that makes this specific geological feature one of the most efficient type of soil in earthen architecture (Arias et al. 1995; Bellanca et al. 1996; Randazzo et al. 2016).

SEM analysis supported the FT-IR preliminary observations, as evidence of conflagration was clearly visible in the signs of vitrification recorded in the samples' microstructure (i.e., bloated pores) (Figure 3a). The wide range-diversity of heat effects on the mudbricks' structure indicated that they were subjected to a variety of temperatures, even when they belonged to the same architectural structure. This supported the hypothesis that mudbricks were not fired on purpose, but that their firing was linked to several accidental conflagrations, which matched the evidence of destruction recovered from multiple Minoan sites (Driessen and MacDonald 2000; Warren 1985).

The SEM also showed the prevalence of illite-muscovite clay in most mudbrick samples, indicating a mixture of different sediment sources as main raw material for production. This analysis revealed the additional presence of various inclusions, such as calcite and dolomites, broadly available across the island, and eventually grog and shell fragments, which were mixed during manufacturing to reduce shrinking, but could also influence the bricks' performance and long-term preservation (Devolder and Lorenzon 2019; Oskouei et al. 2017) (Figure 3b).

Additionally, SEM observations highlighted other important characteristics of Minoan mudbricks that offered insights into their material performance and preservation over time. First, SEM

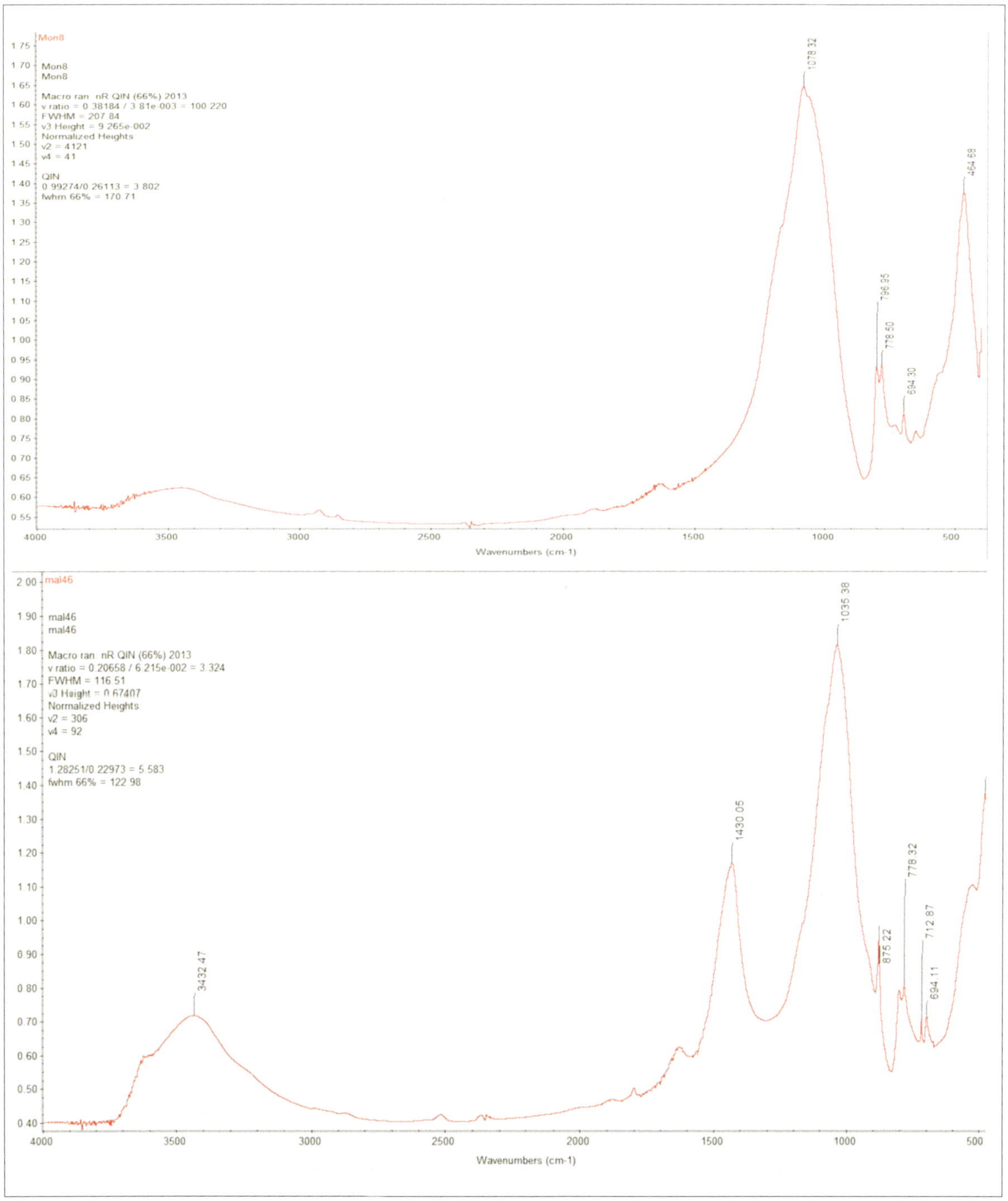

Figure 2. FT-IR spectra of samples MO8 (top) and MA46 (bottom), showing clay and calcite peaks.

microphotographs revealed the extensive presence of vegetal temper within the mudbrick matrix. Vegetal temper is the most common tempering agent in earthen construction, and its presence was visible in the elongated voids of the microstructure and some limited amount of fibrous materials that survived a low-temperature conflagration as charred material (Lorenzon 2017; Marinova et al. 2011) (Figure 3d).

The widespread occurrence of vegetal temper in Minoan mudbricks increased the porosity of the microstructure, affecting the bricks' bulk density. The use of mudbricks with different bulk density was recorded even within the same building (i.e., Malia Palace). The analysis of diachronic mudbrick samples showed a higher porosity in the EBA and initial MBA samples (Pre- and Protopalatial samples, e.g. Gournia and Malia samples; Protopalatial samples in Monastiraki), progressively

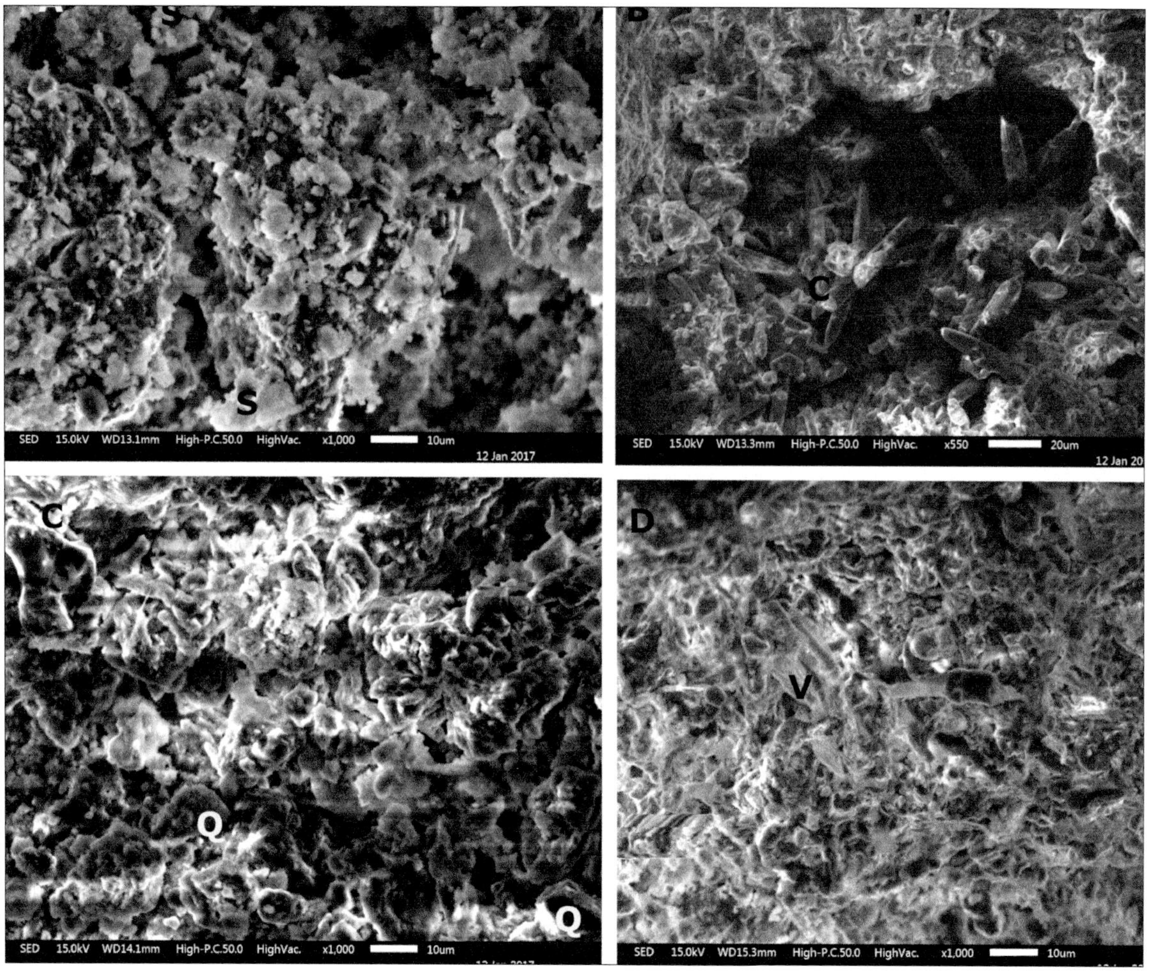

Figure 3. SEM photomicrograph showing microstructure characteristics: 3a. structure that show evidence of vitrification (S) (Sample GOU14); 3b. Calcite crystals (C) (Sample MA5); 3c. The sub-angular quartz grains (Q), standing out against the clay microstructure (sample MA50); 3d. Vegetal temper traces in the microstructure (V) (sample MA25).

decreasing during the MBA and LBA (Neopalatial samples, e.g. Gournia and Malia samples). This reflects the use of diverse vegetal temper and manufacturing practices over time, especially the shift from the use of chaff to the use of seagrasses as main vegetal temper, which affected the bricks' porosity and density (Guest-Papamanoli 1978; Lorenzon 2017).

Moreover, in a limited number of samples from Malia and Gournia, SEM indicated the presence of coarse sub-angular sand grains, demonstrating that marine sediments were also used as natural additive alongside calcite and grog as a degreaser (Figure 3c).

At the structural level, all the samples showed the existence of an extensive network of micro-fissures, that originated naturally during the drying process. The presence of more numerous and wider fissures in the samples dated to the MBA and LBA could have been caused by a high water-retention and humidity variation within these mudbricks (Figure 4). The analysis of Minoan constructions indicated that often MBA and LBA mudbricks were laid into courses when not completely dried, which caused them to spall and deform over time (Devolder 2009; Lorenzon 2017). This latter aspect deeply affected the mudbricks' long-term preservation as evapotranspiration and salt efflorescence linked to water exchange within the bricks permeated and filtered into the micro-fissures, loosening the tight Terra Rossa microstructure. This phenomenon progressively reduced the mudbricks' structural cohesion, leading to crumbling and loss of material (Camerini et al. 2019; Devolder and Lorenzon 2019; Elert et al. 2008).

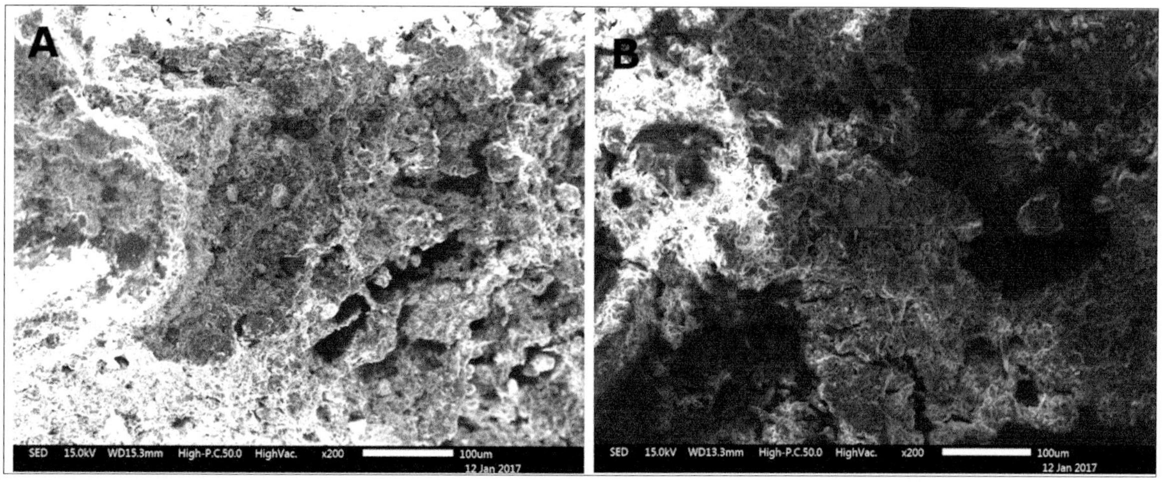


Figure 4. SEM photomicrograph showing fissures and high porosity in the mudbricks' microstructure (samples MO9 and MA5).

### 3 Preliminary Observations

The combined results of FT-IR and SEM clearly indicate that Minoan earthen materials recovered through excavation have in the almost absolute majority of cases undergone a conflagration. Furthermore, the results underline specific patterns of Minoan mudbrick manufacture and performance. For instance, in the manufacturing process the inclusion of various additives was used to produce a more plastic recipe. This includes not only employing vegetal temper as fibrous material to improve the mudbricks' tensile strength, but also mineral additives, such as calcite, sand and grog, to function as stabilizers in order to increase the bricks' mechanical strength, and likely improve their long-term performance (Liberotti and Quaresima 2010; Lorenzon and Iacovou 2019; Nodarou et al. 2008). This paper also emphasizes some of the main structural characteristics of Minoan mudbricks:

- Conflagration deeply impacted Minoan mudbrick clay structure. The traces are visible at the macroscale through burn marks on the surface, and at the microscale through the vitrification of the silicates.
- Both FT-IR and SEM results concurr in determining that mudbricks were heated at extremely diverse temperatures, supporting the hypothesis of accidental fire events that accelerated the mudbricks' collapse.
- The extensive use of diverse vegetal temper affected the mudbrick microstructure by increasing and/or decreasing its porosity and creating diverse bulk densities, depending on the type of vegetal temper used (i.e., chaff vs seagrasses).
- Wall construction with mudbricks that retained a residual water component increased the incidence of wider structural fissuring and led to increased material deterioration.

This study presents multiple insights on the ways the analytical investigation of earthen materials can become a multifaceted instrument to better understand the buildings' preservation history and therefore the causes of their decay (Friesem et al. 2014; Devolder and Lorenzon 2019; Lorenzon 2017).

Although the presence of earthen architecture is often discussed in the archaeological literature of Bronze Age Crete, the material has been significantly undervalued as a source of both environmental and archaeological data. This examination, together with other recent academic studies (Devolder 2009; Nodarou et al. 2008), aims to counterbalance the previous academic narrative, illustrating the relevance of geoarchaeological analysis of earthen constructions not only to better understand the relationship between human and natural environments, but also to grasp the eventual causes of the collapse and decay of buildings.

## Acknowledgements

The author wishes to thank the directors of the three sites, Dr. Kanta, Prof. Watrous, Prof. Poursat, Dr. Pomadère and Dr. Devolder for permission to sample and study the materials. Special thanks to Takis Karakanas, Eleni Nodarou, Vassili Kilikoglou, Maud Devolder, Steve Weiner and Elisabetta Boaretto for their willingness to discuss the results of the methodology with me. Sampling permits were provided by the Ephorate of Antiquities of Heraklion, Ephorate of Antiquities of Rethymnon and the Ephorate of Antiquities of Lasithi and the Greek Ministry of Culture. This study also benefitted from the support of the University of Edinburgh, the Wiener Laboratory for Archaeological Science, the American School of Classical Studies at Athens (ASCSA), the British School at Athens (BSA) and the École Française d'Athènes (Efa).

## Bibliography

Arias, M., Barral, M.T. and Diaz-Fierros, F. (1995) 'Effects of iron and aluminium oxides on the colloidal and surface properties of kaolin', *Clays and Clay Minerals*, 43(4): 406-416.

Bellanca, A., Hauser, S., Neri, R. and Palumbo, B. (1996) 'Mineralogy and geochemistry of Terra Rossa soils, western Sicily: insights into heavy metal fractionation and mobility', *Science of the Total Environment*, 193(1): 57-67.

Berna, F., Behar, A., Shahack-Gross, R., Berg, J., Boaretto, E., Gilboa, A., Sharon, I., Shalev, S., Shilstein, S., Yahalom-Mack, N. and Zorn, J.R. (2007) 'Sediments exposed to high temperatures: reconstructing pyrotechnological processes in Late Bronze and Iron Age Strata at Tel Dor (Israel)', *Journal of Archaeological Science*, 34(3): 358-373.

Bietak, M. (2014) 'Radiocarbon and the date of the Thera eruption', *Antiquity* 88(339): 277-282.

Bradfer-Burdet, I. and Pomadère, M. (2011) 'Δβ at Malia: Two Houses or One Large Complex?', *Hesperia Supplements*, 44: 99-108.

Camerini, R., Chelazzi, D., Giorgi, R. and Baglioni, P. (2019) 'Hybrid nano-composites for the consolidation of earthen masonry', *Journal of colloid and interface science* 539: 504-515.

Devolder, M. (2009) 'From the ground up: Earth in Minoan construction techniques. The case of Building 5 at Palaikastro', *Aegean Archaeology*, 8: 65-80.

Devolder, M. and Lorenzon, M. (2019) 'Minoan Master Builders? A diachronic study of mudbrick architecture in the Bronze Age Palace at Malia', *Bulletin de correspondance hellénique*, 143(1): 63-123.

Driessen, J. and MacDonald, C.F. (2000) 'The eruption of the Santorini volcano and its effects on Minoan Crete', *Geological Society, London, Special Publications*, 171(1): 81-93.

Elert, K., Sebastián, E., Valverde, I. and Rodriguez-Navarro, C. (2008) 'Alkaline treatment of clay minerals from the Alhambra Formation: Implications for the conservation of earthen architecture', *Applied Clay Science*, 39(3-4): 122-132.

Forget, M.C., Regev, L., Friesem, D.E. and Shahack-Gross, R. (2015) 'Physical and mineralogical properties of experimentally heated chaff-tempered mud bricks: Implications for reconstruction of environmental factors influencing the appearance of mud bricks in archaeological conflagration events', *Journal of Archaeological Science: Reports*, 2: 80-93.

Friesem, D.E., Karkanas:, Tsartsidou, G. and Shahack-Gross, R. (2014) 'Sedimentary processes involved in mud brick degradation in temperate environments: a micromorphological approach in an ethnoarchaeological context in northern Greece', *Journal of Archaeological Science*, 41: 556-567.

Goodman-Elgar, M. (2008) 'The devolution of mudbrick: ethnoarchaeology of abandoned earthen dwellings in the Bolivian Andes', *Journal of Archaeological Science*, 35(12): 3057-3071.

Guest-Papamanoli, A. (1978) 'L'emploi de la brique crue dans le domaine égéen à l'époque néolithique et à l'âge du bronze', *Bulletin de correspondance hellénique*, 102(1): 3-24.

Hawes, H.B., Williams, B.A., Seager, R.B. and Hall, E.H. (1908) *Gournia, Vasiliki, and other prehistoric sites on the Isthmus of Hierapetra, Crete: excavations of the Wells-Houston-Cramp Expeditions 1901 1903 1904*. Philadelphia: INSTAP.

Karkanas, P. and Van de Moortel, A. (2014) 'Micromorphological analysis of sediments at the Bronze Age site of Mitrou, central Greece: patterns of floor construction and maintenance', *Journal of Archaeological Science*, 43: 198-213.

Kontogiorgos, D. (2010) 'Tracing the difference: a geoarchaeological approach to the formation of a Neolithic tell/extended site in Greece', *Open Anthropology Journal*, 3: 148-152.

Jerome, P.S. (1993) 'Analysis of bronze age mudbricks from Palaikastro, Crete' in *Conferência internacional sobre o estudo e conservaçao da arquitectura de terra (7a). Silves (Portugal), 24 a 29 de Outubro 1993.* Lisboa: Direcção Geral Dos Edifícios e Monumentos Nacionais: 381-386.

Liberotti, G., and Quaresima, R. (2010) 'Building materials in the 4th and early 3rd millennium monumental architecture at Arslantepe: mudbricks and plaster' in Frangipane, M. (ed.), *Economic centralization in Formative States. The archaeological reconstruction of the economic system the 4th millennium Arslantepe.* Roma: Università La Sapienza: 73-80.

Lorenzon, M. (2017) Earthen architecture in Bronze Age Crete: from raw materials to construction (Doctoral dissertation). University of Edinburgh.

Lorenzon, M. and Iacovou, M. (2019) 'The Palaepaphos-Laona rampart. A pilot study on earthen architecture and construction technology in Cyprus', *Journal of Archaeological Science: Reports*, 23: 348-361.

Love, S. (2012) 'The Geoarchaeology of Mudbricks in Architecture: A Methodological Study from Çatalhöyük, Turkey', *Geoarchaeology*, 27(2): 140-156.

Love, S. (2017) 'Field Methods for the Analysis of Mud Brick Architecture', *Journal of Field Archaeology*, 42(4): 351-363.

Manning, S.W., Höflmayer, F., Moeller, N., Dee, M.W., Ramsey, C.B., Fleitmann, D., Higham, T., Kutschera, W. and Wild, E.M. (2014) 'Dating the Thera (Santorini) eruption: archaeological and scientific evidence supporting a high chronology', *Antiquity*, 88(342): 1164-1179.

Marinova, E., Van Loon, G., De Meyer, M., and Willems, H. (2011) 'Plant economy and land use in Middle Egypt during the late Antique/early islamic period: archaeobotanical analysis of mud bricks and mud plasters from the area of Dayr al-Barshā', *Reports in African Archaeology*, 3: 120-136.

Marocchi, M., Bargossi, G.M., Gasparotto, G. and Dondi, M. (2010) 'Vitrification of basalt orthostats and mud building components from Tilmen Höyük (south-eastern Turkey): an experimental and geoarchaeological approach', *Journal of Archaeological Science*, 37(3): 488-498.

McIntosh, R.J. (1974) 'Archaeology and mud wall decay in a West African village', *World Archaeology*, 6(2): 154-171.

Nodarou, E., Frederick, C. and Hein, A. (2008) 'Another (mud) brick in the wall: scientific analysis of Bronze Age earthen construction materials from East Crete', *Journal of Archaeological Science*, 35(11): 2997-3015.

Oskouei, A.V., Afzali, M. and Madadipour, M. (2017) 'Experimental investigation on mud bricks reinforced with natural additives under compressive and tensile tests', *Construction and Building Materials*, 142: 137-147.

Randazzo, L., Montana, G., Hein, A., Castiglia, A., Rodonò, G. and Donato, D.I. (2016) 'Moisture absorption, thermal conductivity and noise mitigation of clay based plasters: The influence of mineralogical and textural characteristics', *Applied Clay Science*, 132: 498-507.

Shaw, J.W. (2009) *Minoan architecture: materials and techniques.* Series Università di Catania. Centro di archeologia cretese. Padova: Bottega d'Erasmo.

Singh, M., Kumar, S.V. and Waghmare, S. (2016) 'An analytical and geoarchaeological approach on earthen plaster of 2nd century BCE Karle caves of Western Deccan, India', *Journal of Archaeological Science: Reports*, 9: 522-535.

Warren, P.M. (1985) 'Minoan palaces', *Scientific American*, 253(1): 94-103.

Warren, P. (2010). 'The date of the Late Bronze Age eruption of Santorini on the basis of the historical chronology', *Pasiphae*, 4(4): 67-72.

Weiner, S. (2010) *Microarchaeology: beyond the visible archaeological record.* Cambridge/New York: Cambridge University Press.

Zhang, L., Yang, L., Jelle, B.P., Wang, Y. and Gustavsen, A. (2018) 'Hygrothermal properties of compressed earthen bricks', *Construction and Building Materials*, 162: 576-583.

# Chapter 6

# The construction techniques in the middle basin of the Douro between the 3rd and 2nd millennium BCE

Héctor Juan Fonseca de la Torre
Universidad deValladolid
hectorjuan.fonseca@uva.es

**Abstract**

The domestic architecture of the first metallurgical societies of the Northern Spanish Plateau have become of interest until recently, due to the lack of remains in primary context such as postholes, foundations or accumulated rubble. This absence is perceived as a consequence of the constant exploitation of the land, especially since the introduction of agricultural machinery. The main evidence of these huts is the clay remains found accumulated inside holes, which were discarded by archaeologists, considering that no information could be obtained of them. To refute this belief, the remains of construction clay from several sites in the Duero Basin were subjected, basically, to macroscopic analysis, obtaining much information about the construction technique, some ideas about the function of these structures and the detection of possible practices throughout the period studied.

**Keywords:** Copper Age, Bronze Age, Northern Spanish Plateau, Earthen architecture

**Résumé**

L'architecture domestique des premières sociétés métallurgiques du Plateau du nord de l'Espagne n'a suscité aucun intérêt jusqu'à récemment, en raison de l'absence de vestiges en position originale tels que des trous de poteaux, des fondations ou des effondrements. Cette absence est perçue comme une conséquence du travail constant du sol, en particulier depuis l'introduction des machines agricoles. L'information principale de ces maisons provient des accumulations de restes d'argile de construction trouvés à l'intérieur de trous, qui ont jusqu'ici été ignorés par les archéologues, considérant qu'aucune information ne peut en être obtenue. Pour réfuter cette idée, les restes d'argile de construction de plusieurs gisements du bassin du Duero ont été soumis à une analyse essentiellement macroscopique, révélant de nombreuses informations sur la technique de construction, quelques idées sur la fonction de ces structures et la détection de possibles pratiques tout au long de la période étudiée.

**Mots-clés:** Chalcolithique, Âge du Bronze, Plateau Nord de l´Espagne, architecture en terre

## 1 Introduction

Since the beginning of the archaeological research in the Douro Valley, the prehistoric domestic structures were difficult to identify and study due to a combination of adversities. For a better comprehension of those constraining factors we should know the geography of the area of study.

The Northern Spanish Plateau has a quite homogeneous landscape across the Douro Basin. It is a sedimentary basin, mostly covered by a clayey subsoil and completely surrounded by mountainous formations except to the west. In their central area, the presence of calcareous moorlands determined the settlement patterns of the prehistoric communities (Figure 1). Nowadays, the land is occupied principally by agricultural fields but, in the 3rd-2nd millennium BCE the landscape was very different. The palynological studies confirms the abundance of woods (Delibes et al. 2015), especially in the moorlands. The important presence of forests is documented until the 19th century, when they were cut down in order to create more farmlands (Alario and Guerra 2011). With the disappearance of the woods, the majority of streams that crossed the countryside dried out. Therefore, in prehistoric times the water resources were more abundant than today. The wide range of available resources made the plateau a favourable place for the settlement of the first

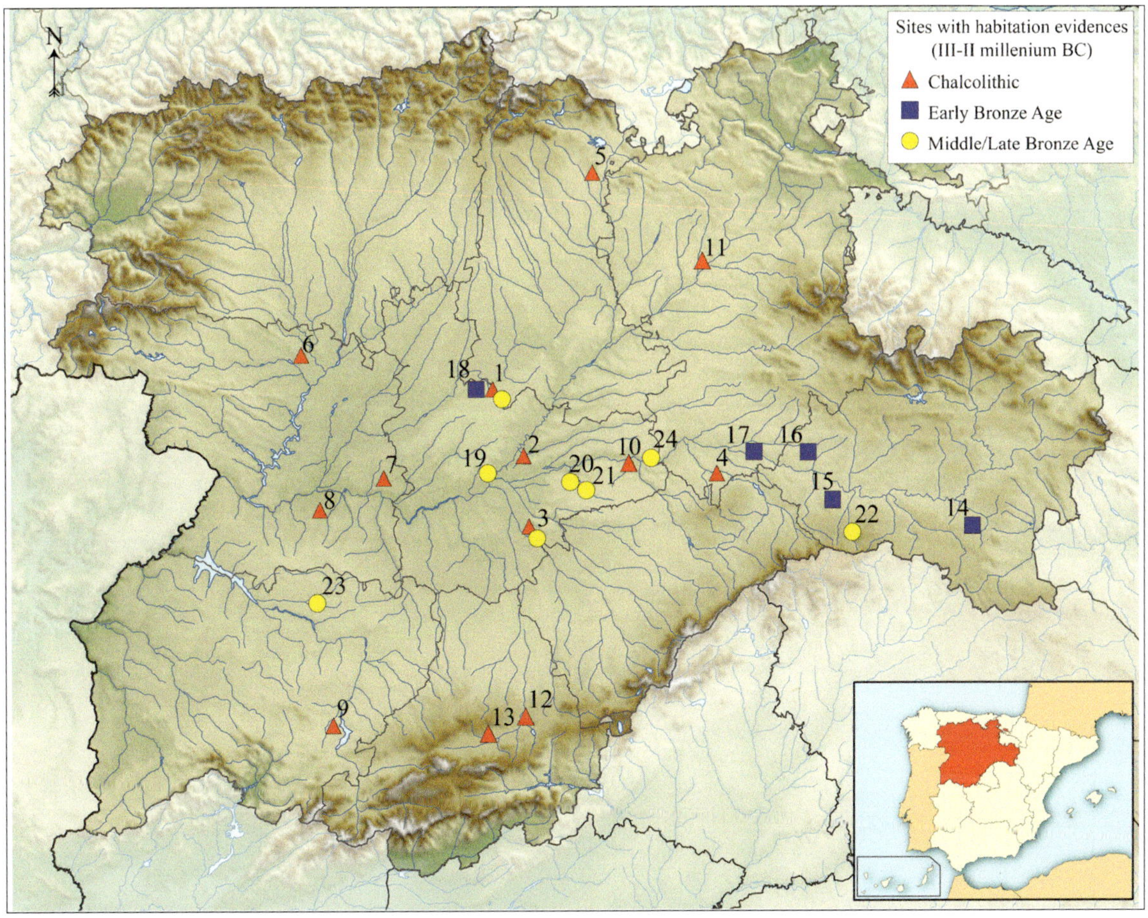


Figure 1. Map of the sites mentioned in the text. Chalcolithic: 1. Casetón de la Era. 2. Santa Cruz III. 3. Prado Esteban. 4. Los Cenizales. 5. Los Doce Cantos. 6. Los Bajos I-II. 7. Las Peñas. 8. Las Pozas. 9. La Viña de Esteban García. 10. Pico del Castro. 11. Los Rompizales. 12. Aldeagordillo. 13. Los Itueros. Early Bronze Age: 14. Parpantique. 15. Los Torojones. 16. Pico de los Cotorros. 17. Pico Romero. 18. Monasterio de Matallana. Middle/Late Bronze Age: 19. Teso de la Macarroña. 20. La Plaza. 21. El Carrizal. 22. Los Tolmos. 23. El Teso del Cuerno. 24. El Gurugú.

farming communities. Nowadays, the agricultural use of the land is one of the main economic activities of region.

Precisely, the uninterrupted use of the land for agricultural labours is the main cause of the lack of information about habitat structures and the spatial organization of the settlements. The ploughing has destroyed part of the archaeological levels, especially since the introduction of heavy machinery in the second half of the 20th century. Such mechanical ploughing penetrates even deeper in the substratum and increases the destruction. On the other hand, the settlements placed on the edges of the moorlands were affected by both natural processes of erosion and by the gravel exploitations. The last one provoked the casual finding of a large range of prehistoric sites, unfortunately by the time the archaeologist arrived, the sites were almost completely destroyed (Rodríguez 2007).

Researchers are of the opinion that the plough is the reason of the widespread absence of structures beyond pits and ditched enclosures (Bellido 1996). There are some evidences interpreted as remains of dwellings, like postholes, foundation ditches or recesses made in the land, as well as hardened clay plaques identified as floors and hearths (García 2004-2005). Most of the dwellings found are isolated findings, due not only to the destruction provoked by the plough but also to the reduced surface excavated of many of these sites. However, even in those cases when excavations were made in non-ploughed place like Las Peñas (Villardondiego, Zamora) (Figure 2c), the structures were also completely swept away. Furthermore, evidence of in situ collapse of habitat structures are scarce.

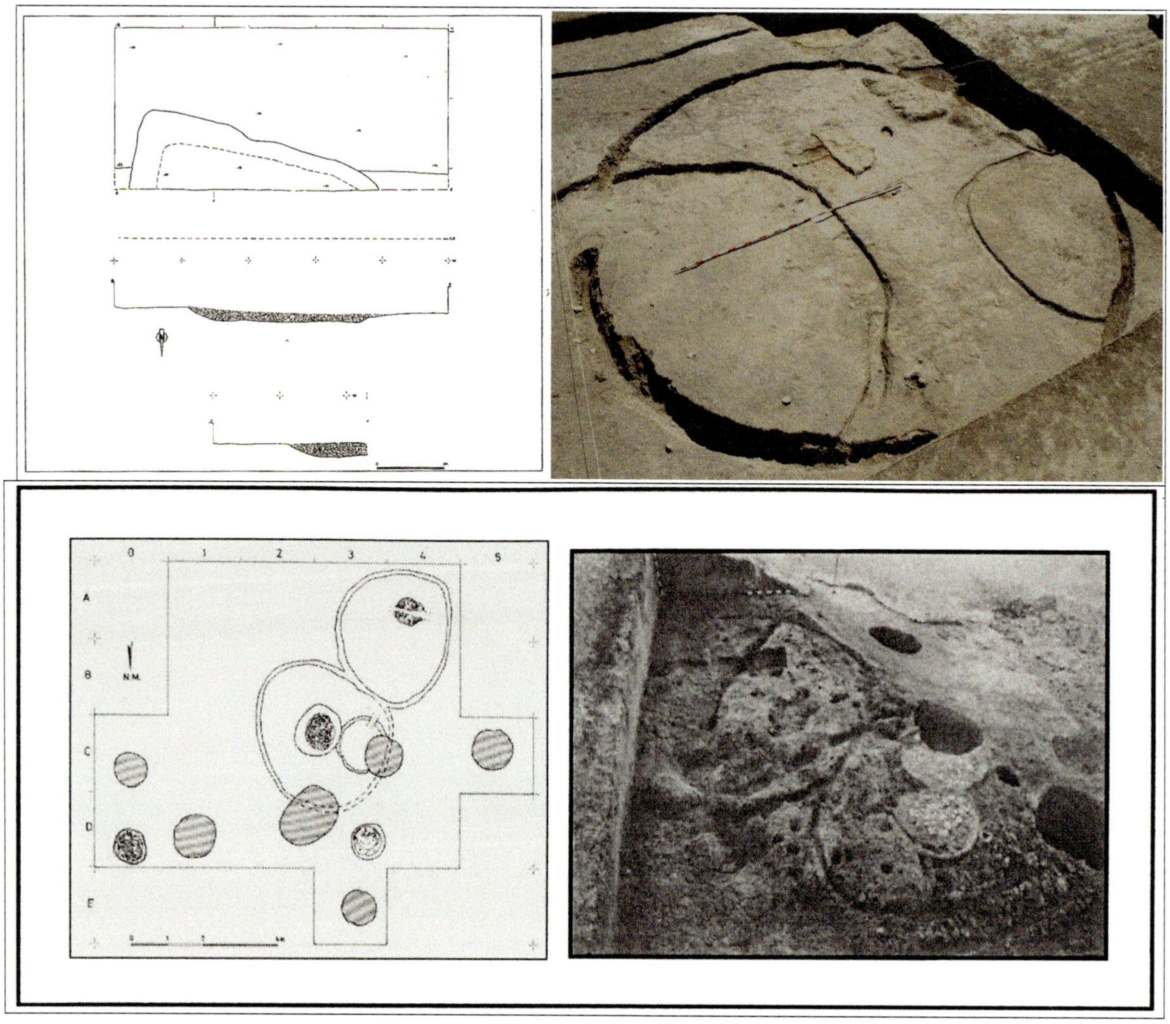

Figure 2. Some examples of structures interpreted as Chalcolithic dwellings: a) Questionable rectangular structure (Los Bajos I-II, Vecilla de Trasmonte, Zamora) (Larrén 1999: 40), b) Four overlapped dwellings foundations and part of an earthen floor (Casetón de la Era, Valladolid) (Delibes et al. 2018: 72), c) Two overlapped dwellings foundations with flanged hearths at the inside (Las Peñas, Zamora) (Delibes et al. 1995: 50).

The main evidence of the existence of habitation structures comes from inside the pits and ditches previously mentioned. These are concentrations of mud fragments hardened by fire that usually have smoothed external surfaces and vegetal imprints. Although mentioned in the excavation reports, these remains are usually discarded in the field or only a small selection of them is collected as a sample. In addition, there is a great terminological confusion in the descriptions of this materials, using terms 'rammed earth', 'adobe' or 'wattle and daub' indistinctly.

That confusion is a symptom of the lack of investigation developed about those findings in the Northern Spanish Plateau. The only precedents are the studies carried out on the sites where clay plaster remains were found chronologically assigned to Protocogotas (Middle Bronze) and Cogotas I (Late Bronze) of La Gravera de Puente Viejo (Mingorría, Ávila) (González and Larrén 1986) and El Teso de la Macarroña (Valladolid) (Arranz et al. 1993). These studies consist of a general description of the remains complemented with the total weight of every concentration found at the site.

The first complex studies that focused on clay plaster remains made in Spain were carried on at the eastern Mediterranean coast. Macroscopic analysis were made at La Vital (Gandía, Valencia) (Gómez 2011) and Cabezo Pardo (San Isidro/Granja de Rocamora, Alicante) (Pastor 2014). Additionally, some archaeometric studies were carried out in order to confirm the presence of whitewashed walls in Mediterranean sites dated in the 3rd millennium BCE (Jover et al. 2016).

In other parts of Spain these studies are still in a very initial phase. This article is a summary of the studies currently carried out in the Douro Valley by the author as part of his PhD project. Not all the settlements with habitation remains registered in the study area were included due to space restrictions, selecting only those with the most significant data for this chapter. Unless otherwise stated, all images are from the author.

## 2 Chalcolithic domestic structures (2900-2200 cal BCE)

The Chalcolithic settlements were placed principally on the Douro Valley, a flat landscape used as crop field over five millennia. As we said before, five thousand years of constant ploughing have destroyed the archaeological habitation level. Only the negative (underground) structures remain unaltered. In some cases, like Prado Esteban (Pedrajas de San Esteban, Valladolid), the unique dwelling found was impossible to date. The reason behind that comes from the absence of any datable evidence inside the postholes from the hut. The site had pits dated to both Copper Age and Late Bronze Age, so is not possible to assign the dwelling to one of those periods (STRATO S.L. 2004).

The main evidences of domestic structures that have been found in this period were postholes, foundation ditches, silos and some mud plaques identified as hearths. There are some possible stone plinths, located all of them at the south of the Plateau, in La Viña de Esteban García (Salvatierra de Tormes, Salamanca) (STRATO S.L. 1992) and Aldeagordillo (Ávila) (Fabián 2006).

In this period, the dwellings' architectural plan was circular or oval in all the cases known and identified as habitation structures, both in the Pre-Bell Beaker and Bell Beaker cultures (Fonseca 2015). There are some rectangular structures that have been identified primarily as dwellings, but these are heavily questioned. The structures unearthed at Los Cenizales (Moradillo de Roa, Burgos) (Sacristán and Elorza 1990: 253) and Los Doce Cantos (Herrera de Pisuerga, Palencia) (Pérez, Fernández and Puertas 1990) were small excavated structures of 8 and 6 m2 respectively and 1 m deep in one of the cases. These structures could have a storage function instead of a residential one. The rectangular structure of Los Bajos II (Vecilla de Trasmonte, Zamora) (Pérez et al. 1993) was a squared shaped recess with no other evidence associated (Figure 2a).

The system of construction of this period can be understood by the great accumulations of clay plaster remains found inside pits in sites such as Las Pozas (Casaseca de las Chanas, Zamora) (Val Recio 1992), Santa Cruz III (Cabezón de Pisuerga, Valladolid) (Delibes et al. 2014: 104-114), La Viña de Esteban García (Salvatierra de Tormes, Salamanca) (Delibes et al. 1997) or El Casetón de la Era (Villalba de los Alcores, Valladolid) (Delibes, Crespo and Rodríguez 2016) among others. These concentrations of clay rubble have been a common find in Chalcolithic sites and suggest a possible intentional dismantling of domestic structures. This may be the reason behind the lack of remains in primary position rather than the effect of the ploughing.

Unfortunately, those vestiges have been commonly discarded or collected only partially. There are few sites where these plaster fragments were systematically gathered. This modus operandi has significantly limited the amount of information obtainable. That is the principal reason behind the significance of El Casetón de la Era as a case study in this matter. The site is a Chalcolithic ditched enclosure reoccupied during the Middle/Late Bronze Age. All clay plaster fragments were recovered and documented during the excavation of the site. These remains appeared not only inside the pits but also inside the ditch, but the largest accumulations came from first ones. Eight foundation trenches were also discovered, concentrated in a zone between the ditch 1 and 2. The ditch 1, the oldest one, sectioned two of them so at least some these dwellings were previous to the construction of the ditched enclosure (Delibes et al. 2018) (Figure 2b). However, none of them had any clay plaster fragment associated.

Figure 3. a) Casetón de la Era: accumulation of daub fragments found inside the pit 70. b) Extraction of the largest accumulation of daub fragments found inside the pit 39 at the site of El Casetón de la Era.

Of the 91 pits excavated in the site, 32 of them had Chalcolithic findings and 78% of these had clay plaster accumulations. Concentrations from pits 70 and 71 were studied in detail through both a macroscopic and instrumental analysis. The details of these studies are collected in several publications (Fonseca et al. 2016; Fonseca et al. 2017) (Figure 3). Now an integral analysis of the clay plaster fragments of the site is being carried out in order to complete the previous ones.

The macroscopic analysis revealed that the morphological characteristics are quite homogeneous between pits. These fragments are also quite similar to those from other sites of the Douro Valley. They have a large size, in some cases superior to 15 cm from side to side, but the size variability is huge. The most relevant feature is the lack or erosion, which demonstrates that the fragments were placed inside the pits immediately after dismantling the structure. This is what preserved these fragments, because unfired clay is quite vulnerable to climatic agents such as wind and rain.

Almost every fragment has vegetal imprints visible on the surface and in the matrix. These imprints can be divided in two major categories: temper and wooden supports. The first group includes those elements that were present into the clay matrix, intentionally as temper (i.e. chaff) or accidentally such as leaves or seeds. Chaff imprints show how the stems were cut to facilitate the mixing. The cut chaff could be a result of the threshing activities documented in the site (Gibaja et al. 2012) (Figure 4c). If the leaves and seeds imprints are in good conditions, the species can be determined. Ilex leaves were identified at El Casetón de la Era (Figure 4b). The second group of wooden imprints are those left by the posts and branches, which compose the dwellings inner structure. The approximate diameter was calculated of both post and branches: posts measure between 12 and 20 cm, while the branches have a diameter of between 2 and 5 cm in most cases. This information then connected with the analysis of the position of the imprints in the matrix and the relations between them, to allow a hypothetical reconstruction of the wooden frame that was the basic structure of these dwellings (Figure 4a and 4d).

This structure consisted on a circle of vertical posts with flexible branches woven between them. The entire structure was covered with several plaster coats in order to isolate the interior. The same constructive technique seems to be used in the rest of the Chalcolithic settlements because their clay plaster fragments present almost the same characteristics.

The circular floor plan revealed by the foundations trenches and postholes is confirmed in the fragments with smoothed surfaces. These are convex or concave depending whether they are from to the outer or inner faces of the wall respectively. The fragments who present flat surfaces came from another zone of the dwellings, such as floors or hearths.

Figure 4. Clay plaster remains from two different sites, Casetón de la Era (CE) and Viña de Esteban García (VEG): a) and d) wood imprints and smoothed surface, b) Quercus ilex leaf imprint, c) wheat spike imprint and fingerprints, e) ridge fragment.

Some hearths could have had a ridge surrounding them. The fragments of these structures are easily recognisable due to their form, which differs from the wall remains. They have a rounded shape and in section it is possible to observe a mud core covered by several clay plaster coatings which gave the desired shape (Figure 4e).

All that information comes from a direct macroscopic analysis carried out on the fragments of a series of sites. Nevertheless, to achieve a deeper knowledge about the constructive (and destructive) processes and confirm the conclusions obtained de visu, it is necessary to make use of a set of instrumental analysis. Currently, El Casetón de la Era is the only site where this kind of analysis has been made on clay plaster fragments, but only from the Chalcolithic occupation (Fonseca et al. 2017). The main contribution of these analyses was the characterization of the main components of the clay matrix. This analysis confirmed a pattern in the temper used in the clay matrix, depending on the part of the dwelling. As an example, hearth fragments had limestone inclusions instead of chaff, which was the main temper used in the walls.

However, the most interesting result of these studies is that the dwellings appear to have been burned down intentionally and the resultant rubble hidden inside a pit. That intention could be demonstrated with the temperatures obtained in the thermal analysis. The two samples analyzed gave quite homogenous temperatures: 530-570 ºC with 800 ºC spikes. A fire of that intensity would not take the walls down as demonstrated by some experiments made with similar dwellings and under similar conditions (Bankoff and Winter 1979; Gheorghiu 2008).

The idea of an intentional concealment of debris is reinforced by the discovery of fragments from hearths and floors. These structures are difficult to dismantle, so it is hard to explain this kind of behaviour from a functional perspective. Similar behaviours have been detected in the Balkans between the 6th and 3rd millennium BCE. The *tell*-sites from this period have occupation levels full of dwellings destroyed by intentional fires (Stevanovic 1997). Tringham (2013: 86-89) created the term 'domithanasia' or house death to express the circumstances of those fires where there is no intention to injure the dwellings' inhabitants.

Now, there is only one case study in the Northern Spanish Plateau referring to an intentional behaviour like that. However, upcoming studies from other sites of the Douro Valley may complete the information in a future and determine if there was an established pattern reflecting a repeated behaviour among those communities.

The Bell Beaker dwellings are extremely rare. There are only two examples in the whole Plateau: El Pico del Castro (Quintanilla de Arriba, Valladolid) (Rodríguez 2005) and Los Rompizales (Quintanadueñas, Burgos) (Alonso 2003). Both consisted of circular structures of which postholes are the only evidence. No daub fragments were found near the postholes or inside any negative structure. The circular form is all that we know about the architecture of these dwellings, which seem to be a formal continuity with respect to the Pre-Bell Beaker. On the other hand, there was a change in the settlement location. During this period there is a noticeable trend to occupy higher places, at the same time as ditched enclosures are abandoned. That change in the settlement pattern matches with a climatic change to a more arid environment (Delibes et al. 2015).

### 3 Early Bronze Age (EBA): the Parpantique horizon (2300-1800 cal. BCE)

The best-known settlements from this phase are concentrated in the Upper Douro Valley. The vast majority of excavations made in other zones consisted of small interventions that provided some negative structures, probably silos, with little amounts of clay plaster fragments. There are only four settlements with dwellings, three of them in the province of Soria: El Parpantique (Ballúncar, Soria), Los Torojones (Morcuera, Soria) and El Pico de los Cotorros (Langa de Duero, Soria) (Fernández 2010); the fourth site (Pico Romero) is located at the south of the province of Burgos, in the village of Santa Cruz de la Salceda (Palomino and Rodríguez 1997). These settlements are on the top of hills that dominate the region and allow visual control of the surroundings (Fonseca and Rodríguez 2017).

Architecturally, the main difference compared with the previous period was the adoption of rectangular floor plan with internal divisions (Fernández and Almeida 2011) (Figure 5d). However, that is an exclusive feature from the eastern zone of the Valley. In the central area, the dwellings remain circular, as the structures found at El Pico Romero demonstrate (Figure 5a).

Another important difference with the Copper Age was the context in which clay-plaster fragments were found. If in the previous period they formed large concentrations inside pits, in this case they were on the surface or associated with the dwellings as it collapsed. One example is the wall fragment from Pico de los Cotorros with imprints of branches that demonstrates the continuity of the wattle and daub technique (Figure 5b).

The external surfaces confirm the presence of straight walls in accordance with the rectangular floor plans. Another aspect of interest lies in the imprints, where traces of planks or rectangular beams are observed for the first time, demonstrating the existence of complex carpentry work (Figure 6c). The abundance of silo cladding fragments identified from their walls and openings, is

Figure 5. Some examples of EBA architectonical remains: a) Post-holes found at Pico Romero, Burgos, b) large piece of daub, Pico de los Cotorros (PC), Soria, c) Daub fragment from Pico Romero, Burgos, d) Hypothetical recreation of a dwelling from El Parpantique, Soria (Fernández and Almeida 2011: 94).

noticeable. In section the technique used is clearly distinguishable, consisting of stacking rolls of mud against the walls of the silo. The surface is then unified by smoothing the rolls and adding a final layer (Figure 6a and b).

About 200 m from the site of El Casetón de la Era is the Cistercian monastery of Santa María de Matallana under whose boundary fence several pits were found, dated to the Early Bronze Age (Crespo, Herrán and Puente 2006). Among these items were several fragments of clay plaster fragments, different from the materials found in the adjacent Copper Age site. These materials are more rounded, in some cases completely shapeless. Among the few recognisable fragments are the remains of pit ridges reminiscent of those found in Soria.

## 4 Middle/Late Bronze Age (M/LBA) (1800-1150 cal BCE)

This period is divided into two phases known as Protocogotas and Cogotas I, based on ceramics decorations: the first one is characterized by incised spikes lines and the second by the use of the boquique technique (geometric incisions incrusted with white paste). The Middle/Late Bronze Age sites are characterised by the horizontal stratigraphy similar to the Copper Age settlements. In fact, many Chalcolithic sites were reoccupied (i.e. El Casetón de la Era) at the same time as the moorland edges were still inhabited, some with defensive structures such as La Plaza (Cogeces del Monte, Valladolid) (Rodríguez and Moral 2007) or El Gurugú (Bocos de Duero, Valladolid) (Rodríguez 1985).

The evidence of buildings is very scarce for this period in the whole valley. This fact sometimes led to the interpretation that these people were itinerant shepherds, whose temporary shelters would leave no mark on the terrain (Jimeno 2001), which contrasts with the intentional destruction of dwellings mentioned above. This lack of evidence of building materials is the main problem for this research. In some cases, finds have been catalogued only as small concentrations of much eroded

Figure 6. a) Silo cladding fragment from El Parpantique (P), Soria. The rolls applied can be distinguished in sectional view. b) a clad pit found at El Parpantique, Soria (Fernández 2010: 80). c) daub fragment with wooden plank/beam imprints from Los Torojones (LT), Soria.

material. This would confirm that residential structures, at the end of their useful life, were given a different treatment than in the Copper Age.

Despite the difficulties mentioned above, there are a number of sites where analyzable material has been recovered. In addition to the Bronze Age reoccupation of El Casetón de la Era, there are the exceptional sites of Los Tolmos (Caracena, Soria) and El Teso del Cuerno (Forfoleda, Salamanca) in which remains of construction mud and dwelling layouts were found. In the first case, there are two adjacent pseudo-rectangular dwellings. Their floor has been levelled by terracing the part of the hill on which they are placed (Figure 7a). There is barely any evidence of postholes, which led their researchers to think of temporary shelters used only at certain times of the year (Jimeno and Fernández 1991).

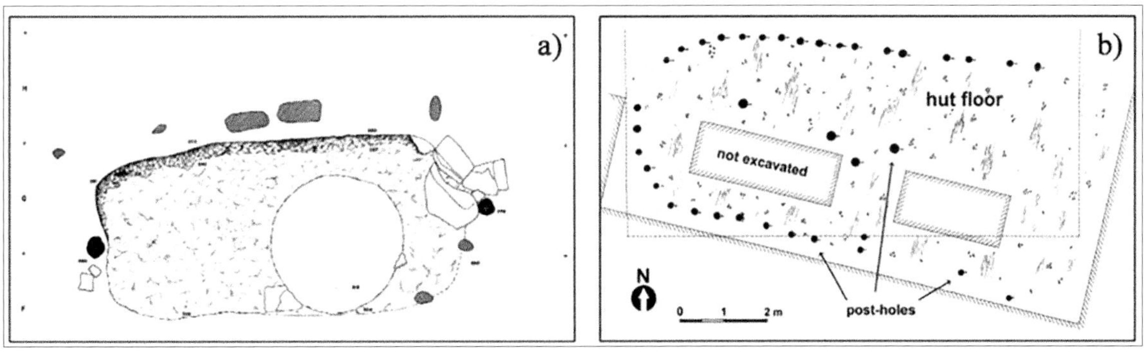


Figure 7. Two examples of M/LBA dwellings: a) Los Tolmos, Soria (Jimeno and Fernández 1991) and b) El Teso del Cuerno, Salamanca (Martín and Jiménez 1989).

On the other hand, El Teso del Cuerno had a set of post-holes that conformed a structure with an apsidal floor plan totally disconnected from the clay plaster fragments, which appear in the interior of several pits dispersed throughout the site (Martín and Jiménez 1989) (Figure 7b).

The fragments of construction mud from this period show a great heterogeneity between sites, but some common characteristics have already been mentioned such as their presence in small concentrations and much more eroded than in previous periods. The presence of straw imprints is also significantly low. However, the greatest regional differences can be seen in the presence of wooden imprints from the structure of the dwelling. In the central and western part of the Duero Valley, this kind of imprints has scarcely been found, which contrasts with the evidence in situ, which consisted of postholes such as El Teso del Cuerno. Los Tolmos, at the eastern area, reports an amount of wood imprints greater than any other contemporary site.

Various fragments from the central and western sites have flattened surfaces, sometimes at 90º angles. Those fragments, and others that do not have smooth faces, are much more compact than those identified as daub. They are reminiscent to some extent of adobe fragments, but the small size of the fragments hinders a more precise interpretation. One of the most obvious is a fragment from El Carrizal (Cogeces del Monte, Valladolid) (Rodríguez Marcos 1990) with two smooth faces forming a right angle; however, its small dimensions (2.6 x 5.7 x 3 cm) exclude it as solid evidence (Figure 8). It is certainly a different construction technique to the previously seen, which uses less straw and wood. The characteristics could fit both adobes (both molded and hand-modelled) or cob (Houben and Guillaud 1994). At the moment, with the fragments found, little more can be said based on a macroscopic analysis.

**5 Discussion**

After reviewing the evidence, we understand the dwellings the Douro Valley better and can now compare the main divergences and continuities over these two millennia.

The type and context of the finds are practically identical, but not the abundance of such discoveries. Although during the Copper Age and the Early Bronze Age the accumulation of material was abundant, the deposition processes were distinct. During the Chalcolithic there is a practice to remove the clay plaster fragments and deposit them in pits, whereas in the Early Bronze Age the fragments are left on the surface where the dwellings collapsed. This last context seems to correspond to the abandonment of the settlement rather than an intentional destruction.

In the Middle/Final Bronze, the fragments were again deposited in pits, but in much lower concentrations and in a state of conservation that presents high degrees of weathering. Thus, during the Copper Age, the deposition of the dwelling remains takes place immediately after the destruction of the structures in what would appear to be a deliberate attempt to erase any trace of the existence of the dwelling, but in the Middle/Final Bronze Age, the removal of the remains

Figure 8. Clay plaster fragment from El Carrizal, Valladolid

is more reminiscent of waste management.

Concerning the construction technique, there was a great homogeneity during the Copper Age and the Early Bronze Age. The study of clay plaster fragments and, in particular, the analysis of their imprints confirms the use of the wattle and daub technique in all the cases studied. The main innovation from one period to another is the evidence of carpentry work as seen in the imprints of beams and/or planks.

There are other construction techniques, but these are used in the manufacture of auxiliary structures, as is the case of cambered hearths and cladding of silos. The remains that comprise this last category are of great interest since, from being non-existent in the previous period, during the Early Bronze Age, they will constitute half of the total amount of recovered clay plaster fragments. There was therefore a desire to guarantee the preservation of the stored elements (presumably grain, as inferred from the seed impressions detected) and to ensure their visibility, since their walls were raised above ground level.

Again, the Medium/Final Bronze Age is a period of change, not only in deposition patterns but also in terms of constructive techniques. Despite the major problems already mentioned, such as the scarcity of fragments and their bad state of conservation, it has been possible to distinguish variants in the morphology of the clay fragments both from one site to another as within the same site. On the one hand, there are light, porous fragments that are similar to the daub fragments from previous periods, but without the imprints of wood, and, on the other hand, there are much more compact fragments, with barely any vegetable material in the matrix (Table 1). As previously mentioned, Los Tolmos seems to be a singular case, as it clearly proves the survival of wattle and daub, though the chaff temper is replaced by leaves.

The case study by excellence is El Casetón de la Era, since, as has been previously stated, it had occupations during practically the entire period studied. It shows how architectural solutions vary independently of the resources available, which theoretically remain constant during this period (Figure 9). However, palynological analyses recorded a change in environmental conditions towards a drier climate, resulting in relative deforestation (López et al. 2014), which could explain the virtual disappearance of wood imprints and the need for alternative constructive solutions.

## 6 Conclusion

In the beginning of this chapter I emphasized the absence of studies focused on constructive clay materials (González and Larrén 1986; Arranz et al. 1993). The main reason for this is that

| Chronological Period | Structures | Remains | Deposition | Conservation | Construction Technique | Destruction |
|---|---|---|---|---|---|---|
| Chalcolithic | Post holes, foundations and hearths | Abundant | Inside pits | Well preserved | Wattle and Daub | Intentional fires? |
| Early Bronze Age | Post holes, Hearths and rubble | Abundant | Inside pits and *in situ* | Well preserved | Wattle and Daub. Stacked rolls for storage coverings. | Fire and/or abandonment |
| Middle and Late Bronze Age | Post holes, hearths, rammed earth floors and rubble | scarce | Inside pits and *in situ* | Rounded and eroded | Great diversity. Almost no presence of wood imprints | Abandonment process |

Tab. 1. Summary of the main features of the archaeological record.

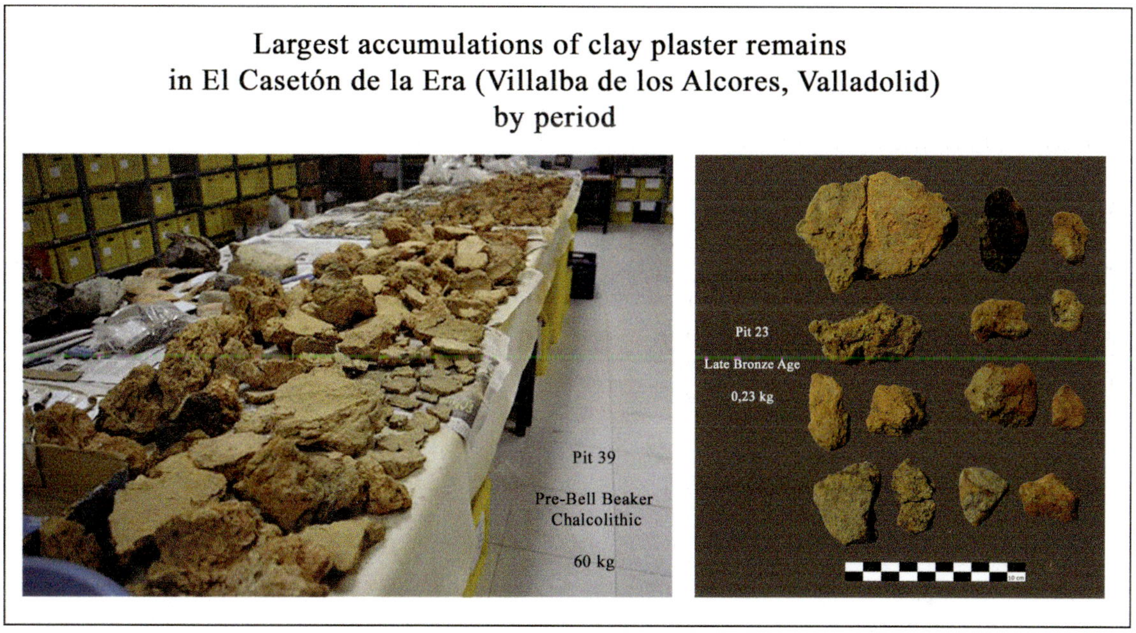


Figure 9. Two clay plaster accumulations found in the same site (Casetón de la Era, Valladolid) from different chronological periods

researchers did not believe these fragments were able to provide information. However, the study condensed in this paper show that it enabled to draw a general panorama of prehistoric domestic architecture in the Douro Valley.

It is possible to describe a certain initial homogeneity during the Copper Age, which is modified in the Early Bronze Age at a formal but not at a technical level: the rectangular floor plan is new, but the wattle and daub technique is maintained. Throughout the Middle/Final Bronze Age the floors plans are diversified, adding to the already known forms the apsidal and pseudo-rectangular layouts. There is also a greater diversity in the morphology of the clay plaster remains, which suggests the population was exploring new architectonical solutions in what could be called an 'experimental' phase (Table 2). Unfortunately, conservation circumstances prevent further study of the construction technique based on macroscopic analysis.

These changes in construction forms and techniques may be caused by two factors. On the one hand, the presence of new groups with a different cultural and technical background could be behind such an abrupt and focused change that took place in the Upper Douro during the Early Bronze Age. But this change

| Chronological Period | Douro Valley | | |
|---|---|---|---|
| | West | Middle | East |
| Chalcolithic | Roundhouses and Wattle and Daub | | |
| EBA | Continuity | | New groups/ideas |
| M/LBA | "Experimental phase" | | |

Tab. 2. Geographical and chronological evolution.

could also be explained from the perspective of an exchange of ideas between different communities, without the need for the settlement of foreign people. On the other hand, the great variability of the Mid/Final Bronze Age and the change in the consumption of certain raw materials seems to be due to a climate change that modified the availability of certain resources and therefore the economic base of those communities. Several of these interpretations should be considered provisional until more consistent data can be obtained for the rest of the Northern Plateau using various analysis techniques, beyond the case study of El Casetón de la Era.

## Bibliography

Alario Trigueros, M. and Guerra Velasco, J. C. (2011) 'Árboles, montes y campos: Los Montes Torozos' in Molinero Hernando, F., Ojeda Rivera, J.F. and Tort i Donada, J. (eds) *Los Paisajes agrarios de España: Caracterización, evolución y tipificación.* Madrid: Ministerio de Medio Ambiente y Medio Rural y Marino: 237-262.

Alonso Fernández, C. (2003) *Excavación arqueológica en el yacimiento Rompizales I (Burgos). Informe inédito.* Burgos: Servicio Territorial de Cultura de la Junta de Castilla y León.

Arranz, J.A., Gómez, A., Sánchez, M. and Bellido, A. (1993) 'El Teso de la Macarroña, un poblado de la Edad del Bronce en la cuenca media del Duero', *Numantia*, 4: 75-93.

Bankoff, H.A. and Winter, F.A. (1979). A House Burning in Serbia. What do burned remains tell an archaeologist? *Archaeology*, 32(5): 8-14.

Bellido Blanco, A. (1996) *Los campos de hoyos. Inicio de la economía agrícola en la submeseta norte.* Valladolid: Universidad de Valladolid.

Crespo Díez, M., Herrán Martínez, J.I. and Puente Aparicio, M.J. (2006) *EL Monasterio Cisterciense de Santa María de Matallana (Villalba de los Alcores, Valladolid).* Valladolid: Diputación Provincial de Valladolid.

Delibes de Castro, G., Abarquero Moras, F.J., Crespo Díez, M., García García, M., Guerra Doce, E., López Sáez, J.A., Pérez Díaz, S. and Rodríguez Marcos, J.A. (2015) 'The archaeological and palynological record of the Northern Plateau of Spain during the second half of the 3rd millennium BC' in Meller, H., Wolfgang Arz, H., Jung, R. and Risch, R. (eds) *2200 BC - A climatic breakdown as a couse for the collapse of the old world.* Tagungen des Landesmuseums für Vorgeschichte Halle, 12, vol. 1. Halle: Landesamt für Denkmalpflege und Archäologie Sachsen-Anhalt, Landesmuseum für Vorgeschichte: 429-448.

Delibes de Castro, G., Benet Jordana, N., Pérez Martín, R. and Zapatero Magdaleno, P. (1997) 'De la tumba dolménica como referente territorial, al poblado estable. Notas sobre el hábitat y las formas de vida de las comunidades megalíticas de la Submeseta Norte' in Rodríguez Casal, A.A. (ed.) *O neolítico atlántico e as orixes do megalitismo: actas do Coloquio Internacional (Santiago de Compostela, 1-6 de abril de 1996).* Santiago de Compostela: Universidad de Santiago de Compostela: 779-808.

Delibes de Castro, G., Crespo Díez, M., Martín Vela, R. and Rodríguez Marcos, J.A. (2018) 'A Ditched Enclosure in the Middle Douro Valley: El Casetón de la Era (Villalba de los Alcores, Valladolid)' in Sastre Blanco, J.C., Rodríguez-Monterrubio, O. and Fuentes Melgar, P. (eds) *Archeology in the River Douro Valley.* Cambridge: Cambridge Scholars Publishing: 61-84.

Delibes de Castro, G., Crespo Díez, M. and Rodríguez Marcos, J.A. (2016) 'Anatomía de un recinto de fosos calcolítico del valle medio del Duero: el Casetón de la Era (Villalba de los Alcores, Valladolid)' in: *Del neolític a l'edat del bronze en el Mediterrani occidental. Estudis en Homenatge a Bernat Martí Oliver.* Valencia: Museo de Prehistoria de Valencia, Diputación de Valencia: 387-401.

Delibes de Castro, G., García García, M., del Olmo Martín, J. and Santiago Pardo, J. (2014) *Recintos de fosos calcolíticos del Valle medio del Duero. Arqueología aérea y espacial.* Studia Archaeologica 100. Valladolid: Universidad de Valladolid.

Delibes de Castro, G., Herrán Martínez, J.I., Santiago Pardo, J. and del Val Recio, J. (1995) 'Evidence of social complexity in the Copper Age of the Northern Meseta' in Lillios, K.T. (ed.) *The origins of complex societies in Late Prehistoric Iberia.* Ann Arbor: International Monographs in Prehistory: 44-63.

Fabián García, F.J. (2006) *El IV y el III Milenio AC en el Valle Amblés (Ávila).* Valladolid: Junta de Castilla y León.

Fernández Moreno, J.J. (2010) El Bronce Antiguo en el oriente de la Submeseta Norte. PhD Thesis. Madrid: Universidad Complutense de Madrid.

Fernández Moreno, J.J. and Almeida Olmedo, J.R. (2011) 'Arquitectura de barro. Reconstrucción gráfica del hábitat del inicio de la Edad del Bronce en el interior peninsular', *Virtual Archaeology Review*, 2(3): 93-97.

Fonseca de la Torre, H.J. (2015) Estructuras de habitación calcolíticas en el Valle Medio del Duero. Un caso de estudio: El Casetón de la Era (Villalba de los Alcores, Valladolid). Trabajo de fin de Máster. Repositorio UCrea. Santander: Universidad de Cantabria.

Fonseca de la Torre, H.J., Crespo Díez, M., Cubas, M. and Sánchez Carro, M.A. (2016) 'Estudio petrográfico de fragmentos de manteado del yacimiento calcolítico precampaniforme de 'El Casetón de la Era' (Villalba de los Alcores, Valladolid)' in Villegas, L. (coord.) *REHABEND 2016 Euro-American Congress. Construction Pathology, Rehabilitation Technology and Heritage Management.* Burgos: Instituto Tecnológico de la Construcción : Universidad de Cantabria : Universidad de Burgos: 145-153.

Fonseca de la Torre, H.J., Crespo Díez, M., Rodríguez Marcos, J.A., Martín Ramos:, Cubas, M. and Sánchez Carro, M.A. (2017) 'Aproximación a la arquitectura del barro en el yacimiento de El Casetón de la Era (Villalba de los Alcores, Valladolid)', in Álvarez Rodríguez, A., Tejedor Rodríguez, C. and García Vázquez, I. (eds) *Arqueología en el Valle del Duero: Del Paleolítico a la Edad Media. Actas de las V Jornadas de Jóvenes Investigadores del Valle del Duero (Valladolid, 12-14 de noviembre de 2015).* Valladolid: Glyphos Publicaciones: 107-124.

Fonseca de la Torre, H.J. and Rodríguez Marcos, J.A. (2017) 'Las técnicas constructivas de la Prehistoria Reciente en el Valle del Duero' in Huerta, S., Fuentes, P. and Gil Crespo I.J. (eds) *Actas del Décimo Congreso Nacional y Segundo Congreso Internacional Hispanoamericano de Historia de la Construcción. Donostia-San Sebastián, 3-7 octubre 2017.* Madrid: Instituto Juan de Herrera, Sociedad Española de Historia de la Construcción: 569-576.

García Barrios, A. (2004-2005) 'El espacio doméstico en la Prehistoria Reciente de la Meseta: el testimonio de las cabañas de la Edad del Cobre en el Valle Medio del Duero', *Lancia*, 6: 59-75.

Gheorghiu, D. (2008) 'Built to be burnt: The Building and combustion of Chalcolithic dwellings in the Lower Danube and East Carpathian areas' in Nikolova, L., Merlini, M. and Comsa, A. (eds) *Circumpontica in Prehistory. Western Pontic Studies. Global gratitude to Eugen Comsa for his 85th birth anniversary.* Oxford: Archaeopress: 55-68.

Gibaja, J.F., Crespo, M., Delibes, G., Fernández, J., Fraile, C., Herrán, J.I., Palomo, A. and Rodríguez, J.A. (2012) 'El uso de trillos durante la Edad del Cobre en la Meseta española. Análisis traceológico de una colección de denticulados de sílex procedentes del 'recinto de fosos' de El Casetón de la Era (Villalba de los Alcores, Valladolid)', *Trabajos de Prehistoria*, 69(1): 133-148.

Gómez Puche, M. (2011) 'El barro cocido' in Pérez Jordà, G., Bernabeu Aubán, J., Carrión Marco, Y., García Puchol, O., Molina Balaguer, L. and Gómez Puche, M. (coords.) *La Vital (Gandia, Valencia): vida y muerte en la desembocadura del Serpis durante el III y el I milenio a.C.* Valencia: Diputación de Valencia, Museo de Prehistoria de Valencia: 229-234.

González-Tablas Sastre, J. and Larrén Izquierdo, H. (1986) 'Un yacimiento del Bronce Medio en Zorita de los Molinos (Mingorría, Ávila)', *Cuadernos Abulenses*, 6: 61-80.

Houben, H. and Guillaud, H. (1994) *Earth Construction: a comprehensive Guide.* London: Intermediate Technology Publications.

Jimeno Martínez, A. (2001) 'El modelo de trashumancia aplicado a la Cultura de Cogotas I' in Ruiz-Gálvez Priego, M. (coord.) *La Edad del Bronce, ¿Primera Edad de Oro de España?.* Barcelona: Crítica/ Arqueología: 139-180.

Jimeno Martínez, A. and Fernández Moreno, J.J. (1991) *Los Tolmos de Caracena (Soria) (campañas 1981 y 1982): aportación al Bronce Medio de la Meseta.* Madrid: Ministerio de Cultura, Dirección General de Bellas Artes y Archivos, Instituto de Conservación y Restauración de Bienes Culturales.

Jover Maestre, F.J., Pastor Quiles, M., Martínez Mira, I. and Vilaplana Ortego, E. (2016) 'El uso de la cal en la construcción durante la Prehistoria reciente: nuevas aportaciones para el levante de la Península Ibérica', *Arqueología de la Arquitectura*, 13, https://doi.org/10.3989/arq.arqt.2016.005 [accessed July 25 2019].

Larrén Izquierdo, H. (coord.) (1999) *Arqueología e infraestructura agraria en el valle del Tera (Zamora).* Madrid: Ministerio de Agricultura, Pesca y Alimentación.

López Sáez, J.A., Abel-Schaad, D., Pérez-Díaz, S., Blanco-González, A., Alba-Sánchez, F., Dorado, M., Ruiz-Zapata, B., Gil-García, M.J., Gómez-González, C. and Franco-Múgica, F. (2014) 'Vegetation history, climate and human impact in the Spanish Central System over the last 9000 years', *Quaternary International*, 353(5): 98-122.

Martín Benito, J.I. and Jiménez González, M.C. (1989) 'En torno a una estructura constructiva en un 'Campo de Hoyos' de la Edad del Bronce de la Meseta española (Forfoleda, Salamanca)', *Zephyrus: Revista de prehistoria y arqueología*, 41-42: 263-282.

Palomino Lázaro, A.L. and Rodríguez Marcos, J.A. (1997) 'Un asentamiento castreño del Bronce Antiguo en la Cuenca del Duero: el Pico Romero en Santa Cruz de la Salceda (Burgos)' in Bueno Ramírez, P. and de Balbín Behrmann, R. (coords.) *II Congreso de Arqueología Peninsular.* Zamora: Fundación Rei Afonso Henriques, vol. 2: 579-590.

Pastor Quiles, M. (2014) 'Cabezo Pardo. Contribución a las formas constructivas de un hábitat argárico a partir del estudio de los elementos de barro' in López Padilla, J.A. (coord.) *Cabezo Pardo (San Isidro/Granja de Rocamora, Alicante). Excavaciones arqueológicas en el yacimiento de la Edad del Bronce.* Memorias Excavaciones Arqueológicas 6. Alicante: MARQ, Museo Arqueológico de Alicante, Diputación de Alicante: 315-329.

Pérez Rodríguez, F.J., Fernández Gutiérrez, J.M. and Puertas Gutiérrez, F. (1990) 'Los Doce Cantos (Herrera de Pisuerga). Un yacimiento calcolítico precampaniforme en el norte de la provincia de Palencia' in Calleja González, M.V. (ed.) *Actas del II Congreso de Historia de Palencia. Palencia: Diputación de Palencia*, vol. I: 95-110.

Pérez Rodríguez, F.J., Sanz García, F.J., Marcos Contreras, G.J., Martín Carbajo, M.A. and Misiego Tejeda, J.C. (1993) 'Algunos aspectos de la Edad del Cobre en el valle medio del río Tera. Intervenciones arqueológicas de urgencia en Vecilla y Colinas de Trasmonte' in *Anuario 1993. Instituto de estudios zamoranos 'Florián de Ocampo'*. Zamora: Diputación Provincial de Zamora: 49-78.

Rodríguez Marcos, J.A. (1985) *Informe de las excavaciones en El Gurugú. Bocos de Duero.* Valladolid: Servicio Territorial de Cultura y Turismo de la Junta de Castilla y León.

Rodríguez Marcos, J.A. (1990) ''El Carrizal' (Cogeces del Monte, Valladolid). Un nuevo yacimiento de facies Proto/Cogotas I', *Numantia*, 4: 61-74.

Rodríguez Marcos, J.A. (2005) 'Una cabaña de época campaniforme: el yacimiento de Pico del Castro (Quintanilla de Arriba, Valladolid)' in *Homenaje al Profesor D. Alberto C. Ibáñez Pérez. Estudios de Historia y Arte.* Burgos: Universidad de Burgos: 81-86.

Rodríguez Marcos, J.A. (2007) *Estudio secuencial de la Edad del Bronce en la Ribera del Duero (provincia de Valladolid).* Valladolid: Junta de Castilla y León.

Rodríguez Marcos, J.A. and Moral del Hoyo, S. (2007) 'Algunos ejemplos de ingeniería prehistórica poco conocidos: tres poblados amurallados del Bronce medio de la sub-Meseta norte' in Arenillas, M., Segura, C., Bueno, F. and Huerta, S. (eds) *Actas del Quinto Congreso Nacional de Historia de la Construcción,* Burgos, 7-9 junio 2007. Madrid: Instituto Juan de Herrera, Sociedad Española de Historia de la Construcción: 775-780.

Sacristán, J.D. and Elorza, J.C. (1990) 'Arqueología Preventiva y de Gestión (1984-1988): Burgos', Numantia. *Arqueología en Castilla y León*, III: 251-257.

Stevanovic, M. (1997) 'The Age of Clay: The Social Dynamics of House Destruction', *Journal of Anthropological Archaeology*, 16(4): 334-395.

STRATO S.L. (1992) *Excavación arqueológica de urgencia en el yacimiento 'Viña de Esteban García', Salvatierra de Tormes, Salamanca.* Salamanca: Servicio Territorial de Cultura, Junta de Castilla y León.

STRATO S.L. (2004) *Actuaciones arqueológicas en los yacimientos de 'Las Almenas', 'Las Cotarrillas', 'Navamboal' y 'Prado Esteban' de la Variante de Iscar y Acondicionamiento de la travesía C-112 de Riaza a Toro por Cuellar y Medina del Campo.* Valladolid: Servicio Territorial de Cultura, Junta de Castilla y León.

Tringham, R. (2013) 'Destruction of Places by Fire. Domicide or Domithanasia' in Driessen, J. (ed.) *Destruction: Archaeological, Philological and Historical Perspectives.* Leuven: Presses Universitaires de Louvain: 89-108.

Val Recio, J. del (1992) 'El yacimiento calcolítico precampaniforme de Las Pozas, en Casaseca de las Chanas, Zamora', *Boletín del Seminario de Estudios de Arte y Arqueología*, 58: 47-63.

## Chapter 7

# The problematic conservation of adobe walls in the open-air site of El Castillar (Mendavia, Navarre, Spain)

Héctor Juan Fonseca de la Torre
Universidad de Valladolid
hectorjuan.fonseca@uva.es, corresponding author

Leyre Arróniz Pamplona
Ayuntamiento de Mendavia
leyrearroniz3@gmail.com

Clara Calvo Hernández
Universidad de Cadizcl
aracalvoh@gmail.com

Laura Cañada Sirvent
Universidad de Alicante
laura.sirvent.canada@gmail.com

Lorena Meana Medio
Independent researcher
lorenameanamedio@gmail.com

Xavier Bayer Rodríguez
Universidad Complutense Madrid
xbayer@ucm.es

Daniel Pérez Legido
Independent researcher
daniel.perez.legido@gmail.com

**Abstract**

The Iron Age settlement of El Castillar (Mendavia, Navarre, Spain) was extensively excavated and studied during the 1970s and 1980s. Afterwards, the site remained open for visitors but without any conservation program. Over the last two decades it was progressively abandoned, until in the summer of 2017 a restoration campaign was carried out in order to conserve, protect and reassess its archaeological importance for the region. In this article analyses the postdepositional processes that affected the adobe structures. As the main purpose of our project is to make El Castillar more accessible and appealing to visitors, the principal difficulty faced for its conservation is that exposing the adobe bricks leads to a rapid weathering of the materials.

**Keywords:** Iron Age, Northern Spain, Earthen architecture, Conservation

**Résumé**

Le village d'El Castillar (Mendavia, Navarre, Espagne) datant de l'Âge de Fer fut largement fouillé et étudié au cours des années 1970 et 1980. Ensuite, le site resta ouvert au public, mais sans aucun programme de conservation. Pendant ces deux dernières décennies, le site fut progressivement abandonné, jusqu'à l'été 2017 où une campagne de restauration à été mise en œuvre en vue de conserver, protéger et réévaluer son importance archéologique pour la région. Dans cet article nous avons l'intention d'établir les processus post-sédimentaires qui ont affecté les structures en adobe. Comme l'objectif principal de notre projet est de rendre El Castillar plus accessible et plus attrayant pour les visiteurs, la principale difficulté recontrée pour la conservation est qu'exposer les blocs d'adobe aux intempéries entraîne une rapide détérioration des matériaux.

**Mots-clés:** Âge du Fer, Espagne septentrionale, adobe, conservation

## 1 Introduction

The protohistorical settlement of El Castillar is located in a north-west hill of the town of Mendavia (Navarre, Spain). The site was extensively excavated during the 1970s and 1980s, becoming the first archaeological site of these characteristics to be studied in the region. After those early excavation campaigns, the site remained open for visitors but after a restoration intervention in 1985 there was no further monitoring. As a consequence, over the last two decades, it was progressively abandoned. In the summer of 2017, a restoration campaign was organized by the Mendavia Town Council, in order to conserve, protect and reassess its archaeological importance for the region.

The excavations revealed several houses. These structures were rectangular and single-room row houses without internal divisions, and they were built along a main street. The rectangular walls of those structures were built on a plinth course of local stone and then raised with both adobe bricks covered with mud plaster and rammed earth, according to the first researches made on the settlement (Castiella 1985). The adobe technique was highly effective and widely used during this period. Other similar settlements such as Cerro de la Cruz (Cortes de Navarra, Navarre) (Maluquer 1954a) and La Hoya (Laguardia, Basque Country) (Llanos 1975) had the same architectonic solutions with very little variation. Only in the case of El Cerro de la Cruz there is no stone plinth, with the wall being completely made of mudbricks from bottom to top.

In addition to the walls, mudbrick was also used to build pottery kilns. These furnaces appeared in at least four different structures. Two of them were in good preservation conditions after their excavation but were left exposed without any protection. As it is the case with the other architectural features of the settlement, all these brick structures needed maintenance after the excavations, but the settlement was abandoned. Therefore, it is possible to analyze the type and intensity of the postdepositional processes after the abandonment in a specific time lapse that affected its conservation. As we will discuss in the following pages, the presence of water, vegetation and animals, transformed the structures in almost unrecognisable accumulations of mud over time. Even the base of the structure, which was the best-preserved section as it was covered by the rubble from the upper course, was affected by those agents, especially by the roots of the plants. In order to avoid having a similar problem in the future, we analyzed the remains to detect the main elements which affect that structure with the intention of establish a restoration protocol.

## 2 The archaeological site

The main objective of 2017 and 2018 campaigns was the recovery and restoration of the housing structures, found in the previous interventions, in order to design a management program for the site so it will not be abandoned again in the future. The main objectives were to improve the accessibility to the site and the understanding of its structures.

As explained in the introduction, the site was in a serious state of abandonment. Vegetation had taken over the old excavation areas, and rabbits and badgers had built dens. As we will discuss later, this situation damaged both the walls of the excavation trenches and the structures.

In addition to these natural problems, it is important to stress the damage made by the use of concrete in order to consolidate the walls. Although commonly used in the 1980s for restoration purposes, concrete has been proven inefficient and sometimes counterproductive when it comes to long-term preservation of archaeological sites (Correia and Guerrero 2018). In this specific case, concrete in many cases caused the stones, used to build the base of the structures, to disintegrate.

Probably due to their good preservation state, most of the adobe structures were left as they were during the 1980s restoration campaign. This fact led to the complete disintegration of the furnace domes and to the partial disintegration of the adobe walls, as we will explain later on.

## 3 The earthen structures

During the excavation campaigns made in the 1970s and 1980s, four earth-made structures were discovered. Three of them were ovens, two of them located in House 1 and another one in House 2. Those ovens were in a good state of preservation and shared the same characteristics: a circular hearth laying on a layer of boulders and covered by a dome with a hole in the front. The whole structure was made of clay shaped by hand. The function of the dome was to concentrate the heat into the oven, and they were probably used as pottery kilns. The adobe wall located in the eastern wall of the House 1, was also found in good condition. By the time it was excavated, it was described as a rammed earth wall (Castiella Rodríguez 1985). That misinterpretation might be due to the final clay coating that hides the adobes from sight. The structure had 6.6 m length was raised on a plinth made with slabs of local stone, commonly known as 'yesón'. The plinth height ranges from 28 to 48 cm and was originally covered by a final clay coating.

In 2017, the general condition of those structures was of complete disintegration. The evaluation made after the removal of the vegetation that had grown on the site over the years confirms that degree of deterioration. The ovens' domes had disappeared completely and the mud cladding and part of top courses of the adobe wall were destroyed. All the structures had the appearance of shapeless mud accumulations (Figure 1).

Figure 1: a) 1982: the adobe wall by the time of its discovery. b) 1985: two ovens from House 1. c) The site at 1991. The ovens were destroyed at that time. d) Photography from spring 2017 (Arroniz et al. 2018). The vegetation completely invaded the structures.

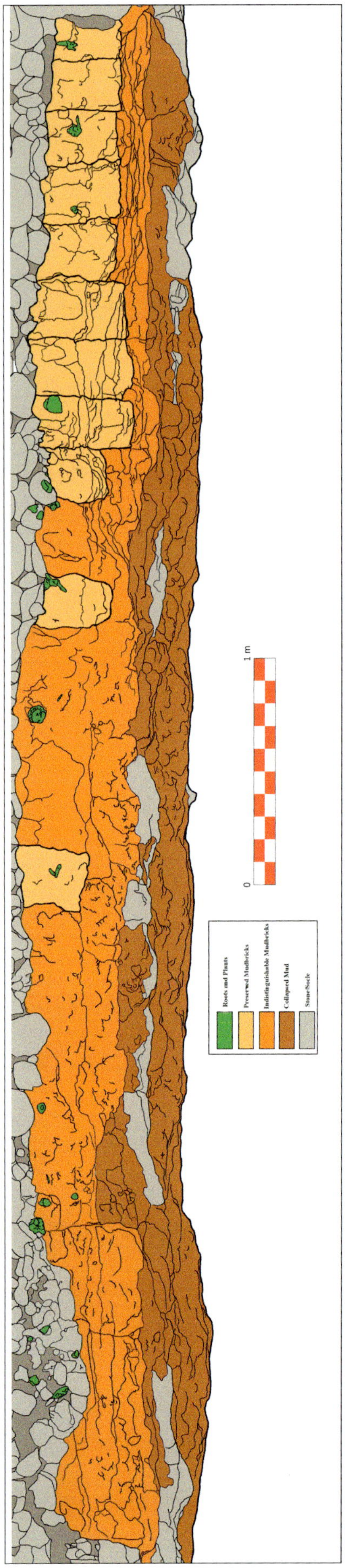


Figure 2. Drawing made over the orthophotos to obtain a clear and precise image of the wall, highlighting the different elements.

## 4 Methodology

The first step in assessing the damage of these structures was to remove completely the eroded layers, to see if it was possible to identify and describe the structures under them. In the case of the wall, once the erosion was removed, we found that despite the bad conservation, some courses and adobe modules could still be distinguished. As it would have been impossible to preserve the structure, we opted to dismantle it systematically and document the process. This way, we could collect the largest amount of information regarding the construction process and evaluate the elements causing the deterioration of the structure. The information about the construction technique can also be used in the future to rebuild the wall for didactic purposes.

Digital photogrammetry was used to document the wall, which was also measured and drawn to scale. After this process, we obtained zenithal and frontal orthophotographs of the three courses we were able to identify, as well as of the stone plinth on which they were placed. Those images were combined with the information collected during fieldwork and, with the 3D model itself, have allowed to identify the different adobe bricks and their measurements (Figure 2). The pottery kilns were completely eroded, so we could not carry out this process on them.

## 5 Results

The wall, when excavated, consisted of four adobe courses. The top course was now completely lost, as well as the wall covering, due to different factors. The complete deterioration of these elements resulted in the formation of a collapsed layer that accumulated the base of the wall into the house.

After cleaning the eroded surface, we were able to identify the adobe bricks of the next course, as well as the maximum width of the wall. However, the conservation of this course is uneven. In the north, it is possible to distinguish a series of nine mudbricks laid on headers. These bricks have quite regular dimensions that denote the use of the same module of 25.5 x 42.5 x 10 cm. In the remaining four meters, this distinction is impossible to make. The wall was almost a formless mass of mud in which we could only partially identify some adobes apart from those described before. Between the best-preserved mudbricks and the total width of the wall given by the stone plinth of the base there was a space of about 25 cm occupied by a mass of disintegrated adobe. This space is the same as the width of the mudbricks conserved, so there was probably a second line of adobe, made with the same mold but laid on stretchers, completing the wall. However, the deficient conservation made it impossible to verify this interpretation.

Despite the irregular conservation of this wall, we can see that some stones used in the restoration of the next house overlaps the mudbricks in both the north and the south sides of the structure. Those were interpreted in the 1980s as part of two stone corners with the rammed earth wall placed between them. Nowadays we can say that the wall consists of a uniform structure with a low stone plinth and adobes disposed over it instead of a rammed earth wall with stone corners. The second course of adobes showed a similar conservation state. In this course, we only have seven consecutive mudbricks. An eighth can be distinguished in the north end, partly covered by a layer of stones that was placed during the previous restoration. We decided not to dismantle this stone layer in an attempt to preserve the modules under it. These modules have dimensions of 25.2 x 40.2 x 10 cm, so they were probably made with the same mold used for the course over it.

The last identified course does not run over the entire plinth, as the southern side is higher. For this reason, they use the first adobe course to level the difference. The mudbricks are arranged on headers, and we can distinguish ten consecutive bricks. Their dimensions are 23.5 x 40.5 x 10 cm, sufficiently similar to the previous so as to say that the same mold size was used. Lastly, we can establish that between the stone plinth and the mudbricks, there is a regularizing mud layer that was also used to mortar the plinth stones in place (Figure 3).

All courses have been affected by roots. They contribute to hinder the identification of the modules as well as the general preservation. It has been impossible to extract even one complete adobe because roots crossed all of them.

The kilns suffered from not only the same erosive agents as the adobe wall but also from others. Surrounding them, as well as in other locations of the settlement, there are badgers' burrows. The last agent that hindered their conservation is human, as plunderers dug on and around them. Because of all that, nowadays we have almost no trace of them (Figure 4).

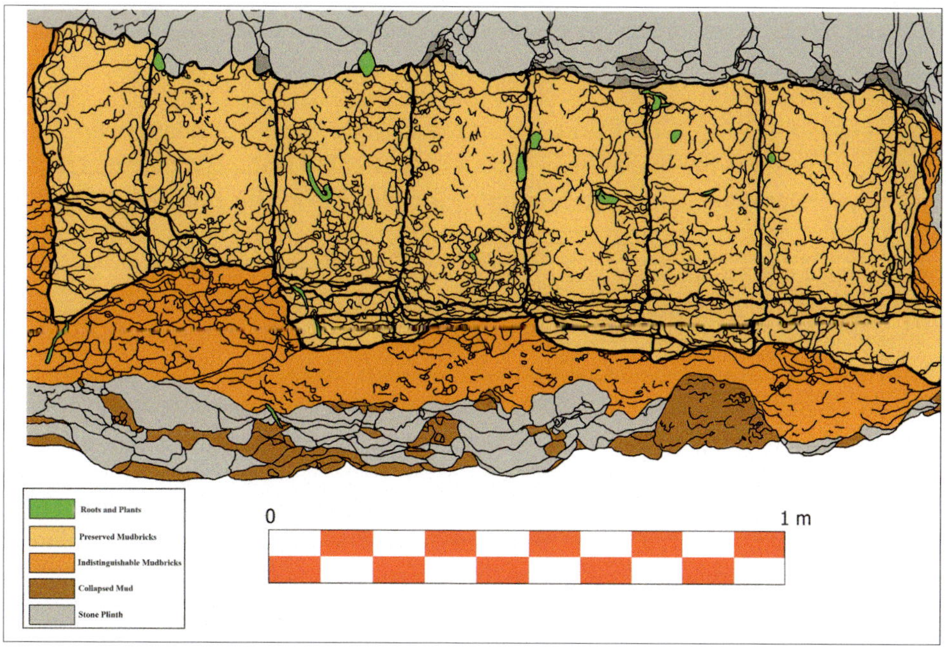


Figure 3. Detail from the second course of mudbricks.

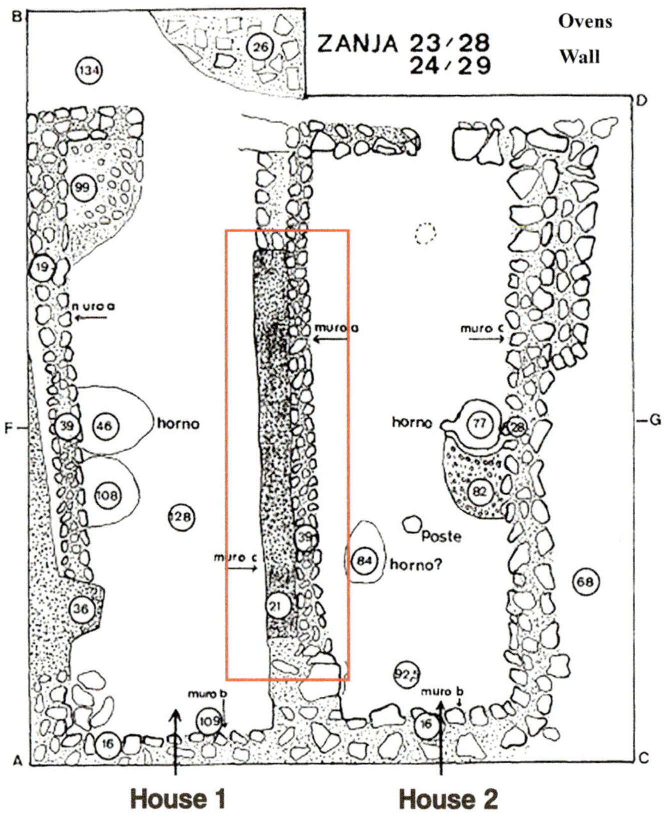


Figure 4. Floor plan made in the eighties with the structures studied marked (Castiella 1985).

## 6 Discussion

Dismantling these deteriorated structures can be considered a proper archaeological excavation by itself, as the objective was to obtain answers to both technological and preservation questions. From a technological point of view, we were interested in knowing the internal structure of the adobe wall, as the closest parallels are from old archaeological interventions, like those of Maluquer de Motes in the archaeological site of Cortes de Navarra in the 1950s. The walls discovered in that settlement were slightly different from ours due to the almost complete absence of stone plinths (Maluquer 1954: 149) but identical in the remaining technical aspects. The results obtained allowed us to discard the first interpretation given to that wall as a rammed earth structure.

As explained previously, we can identify two courses on this wall, an interior one, laid on headers, and an exterior one laid on stretchers. The latter is almost completely lost due to the erosion. The width of the adobes fits with the missing space between the interior course and the maximum width of the stone plinth, so probably it was occupied by mudbricks of the same dimensions in that position.

Thanks to the dismantling of the wall we have been able to document which processes hindered the conservation of these structures, as well as to what extent they caused an impact. These processes are mainly those produced by the action of water and vegetation. They usually happen in combination, generating a negative feedback effect that ends up with the total loss of the structures.

The geographical area of Mendavia has a climate with abundant rain throughout the year. Therefore, water is the main erosive agent that affected these structures. The rain causes leaching on the exterior clay facing, eroding natural conduits on the exposed mudbricks to evacuate the excess water. The subsequent evaporation of water produces new cracks while enlarging existing ones. These cracks increase the erosive action and at the same time they create a favourable environment for vegetation growth (Alva and Chiari 1990: 114-115).

As the plants germinate (especially shrub species as thyme, which is abundant in the surroundings), the roots penetrate into the wall structure, sometimes following the boundary between two bricks, and other times going through them following the cracks. The presence of these roots helps rainwater to penetrate the bricks, contributing to its degradation.

This process is aggravated by the presence of insects, especially annelids, which make holes in the softer parts of the mudbricks. Despite their impact being lower than the other natural factors, the combined result is the almost complete destruction of the structures, within 30 years.

Knowing the agents that erode the mudbrick structures, we can now define the actions needed to mitigate them, as the probability of the appearance of mudbrick structures in future archaeological campaigns is high. All these actions must aim to avoid vegetation growth and water damage (direct rain or capillary absorption).

One of the most extended solutions to preserve this kind of structures is building a permanent roofing over the archaeological structures to protect them from the weather agents. Despite being a long-term conservation proposal, it has major obstacles. First, the high cost of the building, second, the high impact on both the landscape and the archaeological context. Such roofs can also produce erosion of the archaeological soil surrounding them, as water discharged by it always ends up in the same place, and as it can also alter the environmental hygrometry, it still affects the covered structures (Guillaud 2017). However, this solution has been applied in many archaeological sites of various chronologies and characteristics because it is easy to adapt to different environments. The Cerro de San Vicente (Salamanca) (Blanco, Macarro and Alario 2017), the roman villa of La Olmeda (Saldaña, Palencia) or the roman baths of Maliaño (Maliaño, Santander) are a few examples of this solution. In our case, we cannot use this solution, due to the morphology of the hill and the size of the site, as we would not be able to lay the foundations that the structure would require. For this reason, we must choose other kinds of solutions, taking into account everything that we have previously explained.

A well-known option, used especially during the last century and applied mainly to masonry walls, is to apply sacrificial coatings on the top of the wall and earthen renders on the exposed surfaces, acting as a protective layer. Following the reversibility principle, while putting geotextile membranes before these surface coatings, the earthen structures can be protected with the possibility of subsequent removal. Yet because the coatings can also alter the visual relation of archaeological sites their form, height and materials must be carefully and sensibly designed (Correia and Guerrero 2018). Following the rules of Cultural Heritage Restoration, the screed must be clearly distinguishable from the original; preferably, this distinction should be discernible at a close range to limit the visual impact on the visitor (Stubbs 1990).

As indicated above, in this kind of interventions it is very important to choose the materials carefully; in our case, cement was chosen in the 1980s, and the result was completely the opposite of the desired outcome. In the case of adobe structures, the option has often been to build new courses with bricks composed by earth concrete, by mixing cement with soil. Cement must never be present in more than a 10% of the mixture. The result would be a higher strength top course that would protect the remaining wall structure (Alva and Chiari 1990). However, cement is not appropriate because it generates salts and has too high resistance, which makes it incompatible with the original.

Currently the most common practice is to consolidate the structure and isolate it from the main erosive agents, using chemical treatments. Ethyl silicate is usually used in vertical surfaces, but in areas with contrasting temperature changes and salts, it can produce exfoliation ending in the total loss of the material. PVA treatment is also discouraged, since it generates antiperspirant films that trap capillary water within the wall. The most appropriate treatment for in situ consolidation might be acrylic resin (Primal/Acril 33), after moistening the original surface with demineralized water to enhance penetration. This treatment has been applied in the archaeological complex of Chan Chan (12th-15th century CE), Peru. Tests have been carried out, showing that the application of plastic chemical substances, such as Primal AC 33 (acrylic emulsion), it is a positive method that does not present problems of surface exfoliation due to the effects of the moisture contained in the structure (Morales 1983). However, it is important to emphasize that chemical treatments are only used when no other treatment would be effective for the consolidation of the materials, as it is not advisable to use harmful materials.

With all these factors in mind, the most appropriate solution for the preservation of mud structures that are going to remain exposed in situ, is their protection by the construction of at least two sacrificial coatings putting geotextile membranes between the archaeological structure and our intervention. A good documentation of the intervention is fundamental to establish where the new build start sand the archaeological remains end. These interventions require supervision, especially during the first year. During that period of time, it should be checked every two months to see if the restoration stabilizes with the environment. Regular supervision is essential to monitor the state of the structures, replacing the protection courses when necessary. This is the least aggressive intervention and it would allow the visitor a better understanding both archaeological site and of the importance of preservation strategies. As previously mentioned, the ovens are destroyed, so no restoration work can be carried out. Therefore, the possibility of reconstructing one of them for didactic purposes is being considered.

## 7 Conclusion

Adobe structures are of great strength and durability as long as they are taken care of. The abandonment of these structures causes the collapse of the roof and the degradation of the clay facing that covers the mudbricks. In the case of El Castillar, the base of the mudbrick wall is separated from the ground by a stone plinth while the roof of the house protects the top part. After the abandonment of the town, the collapse of the house itself protected the lower section of wall and the pottery furnaces. The abandonment of the site after its excavation in the 1970s and 1980s caused the loss of its structural integrity, since it was left in open air without any cover to protect it from the degradation mechanisms that most affected its upper parts (the most vulnerable areas of mud buildings). The erosion is associated with the rain and wind, that later propitiated the appearance of the other agents mentioned above (Mileto et al. 2017).

The dismantling of the structure, required due to its bad state of preservation, has allowed us to gather valuable information about it. At the archaeological and architectural level, the documentation of the degradation process has allowed us to acquire key knowledge for the future treatment of similar structures that can appear in the site (Arróniz et al. 2018). On the other hand, the documentation of the courses and brick modules of the walls has allowed us to understand the technical complexity of the inhabitants of El Castillar during the last occupation period. The uniformity of the measurements of the mudbricks demonstrates the use of a standardized technique, probably made by specialized artisans. Its placement and the two orientations verify basic architectural knowledge, optimizing both the space and the amount of raw material needed for the construction.

Despite the bad preservation of the structures, previous photographic documentation enables the reconstruction of at least one of the pottery furnaces, in order to offer a more didactic space for visitors.

The loss of the wall and the ovens has allowed us to understand the erosive agents that have been decisive for their destruction. From this knowledge, it has been possible to establish a general guideline for a protection protocol, as our main objective is the preservation and revalorisation of the archaeological site without causing new losses in the future.

## Bibliography

Alva Balderrama, A. and Chiari, G. (1990) 'Protección y conservación de estructuras excavadas de adobe' in Stanley N.P. (ed.) *La conservación en excavaciones arqueológicas.* 2nd edn. Rome: ICCROM: 113-123.

Arróniz Pamplona, L., Meana Medio, L., Sirvent Cañada, L, Fonseca de la Torre, H. J., Pérez Legido, D. and Bayer Rodríguez, X. (2018) *Proyecto de Intervención Arqueológica en el Poblado Protohistórico de El Castillar (Mendavia)*. Memoria de la Campaña de 2017. Pamplona: Gobierno de Navarra (unpublished report).

Blanco González, A., Macarro Alcalde, C. and Alario García, C. (2017) 'La aldea del Hierro Inicial del Cerro de San Vicente (Salamanca, España): Resultados de las excavaciones entre 1990 y 2006 a la luz de algunos debates actuales', *Munibe Antropologia-Arkeologia*, 68: 217-236.

Castiella Rodríguez, A. (1985) 'El Castillar de Mendavia. Poblado Prehistórico', *Trabajos de arqueología Navarra*, 4: 65-143.

Correia, M. and Guerrero, L. (2018) 'Conservation of Earthen Building Materials' in López Varela, S. L. (ed.) *The Encyclopedia of Archaeological Sciences*. Hoboken: John Wiley and Sons, DOI: 10.1002/9781119188230.saseas0117 [accessed July 25 2019]

Guillaud, H. (2017) 'Conserving earthen architecture, a future requirement for a recreated modernity' in Vegas, F. and Mileto, C. (coords.) *The COREMANS Project. Intervention Criteria for Earthen Architecture*. Madrid: Ministerio de Educación, Cultura y Deporte: 134-140.

Llanos Ortiz de Landaluce, A. (1975) 'Excavaciones arqueológicas en el poblado de la Edad del Hierro de la Hoya, Laguardia (Álava)', *Noticiario arqueológico hispánico*, 4: 273-288.

Maluquer de Motes, J. (1954a) 'Los poblados de la Edad del Hierro, de Cortes de Navarra', *Zephyrvs: revista de prehistoria y arqueología*, 5: 1-16.

Maluquer de Motes, J. (1954b) *El yacimiento hallstático de Cortes de Navarra. Estudio crítico-I. Excavaciones en Navarra IV*. Pamplona: Diputación Foral de Navarra, Institución 'Príncipe de Viana'.

Mileto, C., Vegas, F., Vendreil, M. and García Soriano, L. (2017) 'Degradation mechanisms: phenomena and causes' in Vegas, F. and Mileto, C. (coords.) *The COREMANS Project. Intervention Criteria for Earthen Architecture*. Madrid: Ministerio de Educación, Cultura y Deporte: 197-207.

Morales, R. (1983) 'La conservación de estructuras y decoraciones de adobe en Chan Chan', in Truel, J. (ed.) *El adobe. Simposio internacional y curso-taller sobre conservación de adobe: informe final y ponencias principales*. Lima: Proyecto Regional de Patrimonio Cultural y Desarrollo PNUD/UNESCO: 109-117.

Stubbs, J. H. (1990) 'Protección y exhibición de estructuras excavadas' in Stanley N.P. (ed.) *La conservación en excavaciones arqueológicas*. 2nd ed. Rome: ICCROM: 85-102.

# Chapter 8

# Earthen architectural diversity at an early village in the U.S. Southwest

James R. Allison
Department of Anthropology, 800 KMBL Brigham Young University, Provo, Utah 84602, USA
jallison@byu.edu, corresponding author

Joseph A. Bryce
Silver City Museum, 312 West Broadway, Silver City, New Mexico 88061, USA
josephabryce@gmail.com

**Abstract**

Early Pueblo villagers used a variety of earthen construction techniques. We explore these variations in earthen architecture at one early Pueblo village, Alkali Ridge Site 13. These techniques incorporated stone, wood, and other plant material in various ways to modify and strengthen the earthen matrix, and reveal the use of four kinds of buildings. We present a case study, based on two very different research traditions and bodies of theory. The first is the study of technological systems, the second analyzes vernacular architecture. These traditions are focused on the relationships among architecture, culture, and society. This is important because some buildings at Site 13 would have appeared identical in their finished form, but were built using different materials and techniques. Differences in the operational sequence began with the initial steps of gathering raw materials for house construction, and included differences in the ways walls were built and pit structures were roofed.

**Keywords:** Pueblo Culture, Construction techniques, Vernacular architecture, Traditional building systems

**Résumé**

Les villageois de la culture Pueblo ont utilisé diverses techniques de construction en terre. Nous explorons ces variations de l'architecture dans l'un des plus anciens villages Pueblo, Alkali Ridge Site 13. Ces techniques incorporaient la pierre, le bois et d'autres matériaux végétaux de différentes manières pour modifier et renforcer la matrice en terre, et révèlent l'utilisation de quatre types de bâtiments. Nous présentons une étude de cas, basée sur deux traditions de recherche et d'approches théoriques très différentes. La première est l'étude des systèmes techniques, la seconde est l'analyse de l'architecture vernaculaire. Ces traditions sont axées sur les relations entre l'architecture, la culture et la société. Ceci est important, car certains bâtiments du Site 13 semblaient identiques dans leur forme finale, mais furent construits avec de différents matériaux et techniques. Les différences dans la séquence opérationnelle commencent dès les étapes initiales de l'obtention des matières premières pour la construction des maisons et les façons de construire les murs et les toits des structures semi-enterrées.

**Mots-clés:** Culture Pueblo, techniques de construction, architecture vernaculaire, systèmes traditionnels

## 1 Introduction

Early Pueblo villagers in the Southwestern United States used a variety of earthen construction techniques. These techniques incorporated stone, wood, and other plant material in various ways to modify and strengthen the earthen matrix. We explore the variation in earthen architecture at one early Pueblo village, Alkali Ridge Site 13. The Ancestral Pueblo area includes portions of four U.S. states: Arizona, New Mexico, Colorado, and Utah. Site 13 is located near the southeastern corner of Utah, within the Mesa Verde region, an archaeologically defined subdivision of the Ancestral Pueblo world that includes much of southwestern Colorado and southeastern Utah (Figure1). The early village at Site 13 dates to the late 700s, during the early part of what archaeologists refer to as the Pueblo I period. A recent review of the dating suggests that the early Pueblo I village began to be built about 750 CE and was probably abandoned by 780 CE (Allison, n.d.). This was a time of rapid change in the Ancestral Pueblo world, as Pueblo people for the first time began to live in above ground rooms, and simultaneously shifted from a relatively dispersed settlement pattern to

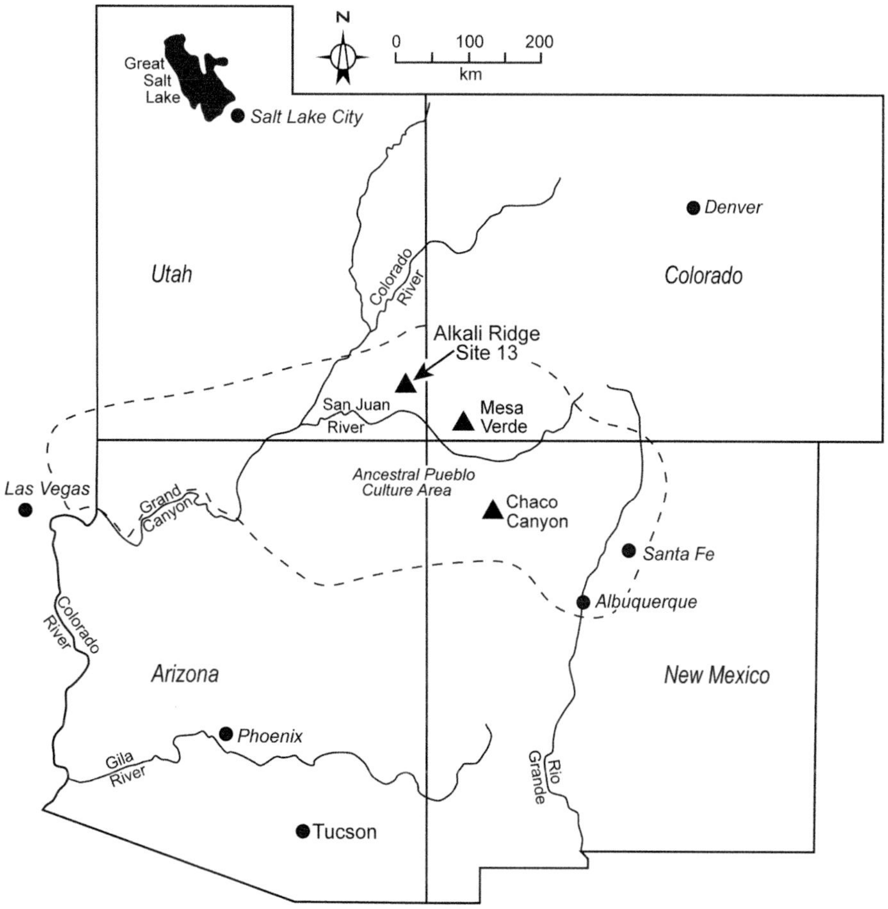


Figure 1. Map of the U.S. Southwest showing the location of Alkali Ridge Site 13, main cities, and well-known archaeological sites. The dashed line shows the approximate boundaries of the Ancestral Pueblo culture area.

larger settlements of up to a few hundred people (Allison et al. 2012; Wilshusen 2018; Wilshusen, Hurst and Chuipka 2012; Wilshusen and Potter 2010).

Site 13 is one of the largest and most extensively excavated Pueblo I villages in the Southwestern United States. Excavations in the 1930s uncovered 118 storage rooms, 11 pit houses, and 25 surface habitation rooms (Brew 1946, Figure 2). We did additional excavations there in 2012 and 2013 that were on a much smaller scale, but our excavations have helped clarify construction characteristics and have recovered architectural material not collected in previous excavations (Figure 2). Much of the site burned, including all the rooms we excavated. Earthen construction materials are preserved in the burned rooms as pieces of fire-hardened daub, and many of these pieces show the impressions of wood beams, other organic materials, or the builder's fingers.

In this paper, we combine observations made during the 1930s excavations (which lack detail but provide a broad overview of variation in construction techniques across the village) with detailed analysis of the recently recovered materials, including fragments of fired earthen construction materials. This allows us to document the technology used to create the earthen architecture at Site 13, and how construction techniques vary across the site.

As we noticed, construction technology varies with structure function. Different techniques were used for surface habitations, storage rooms, residential pit houses, and large communal pit structures. But techniques vary within structure types as well, and this variation likely reflects the fact that early villages formed through the aggregation of numerous small social groups with diverse origins and local traditions.

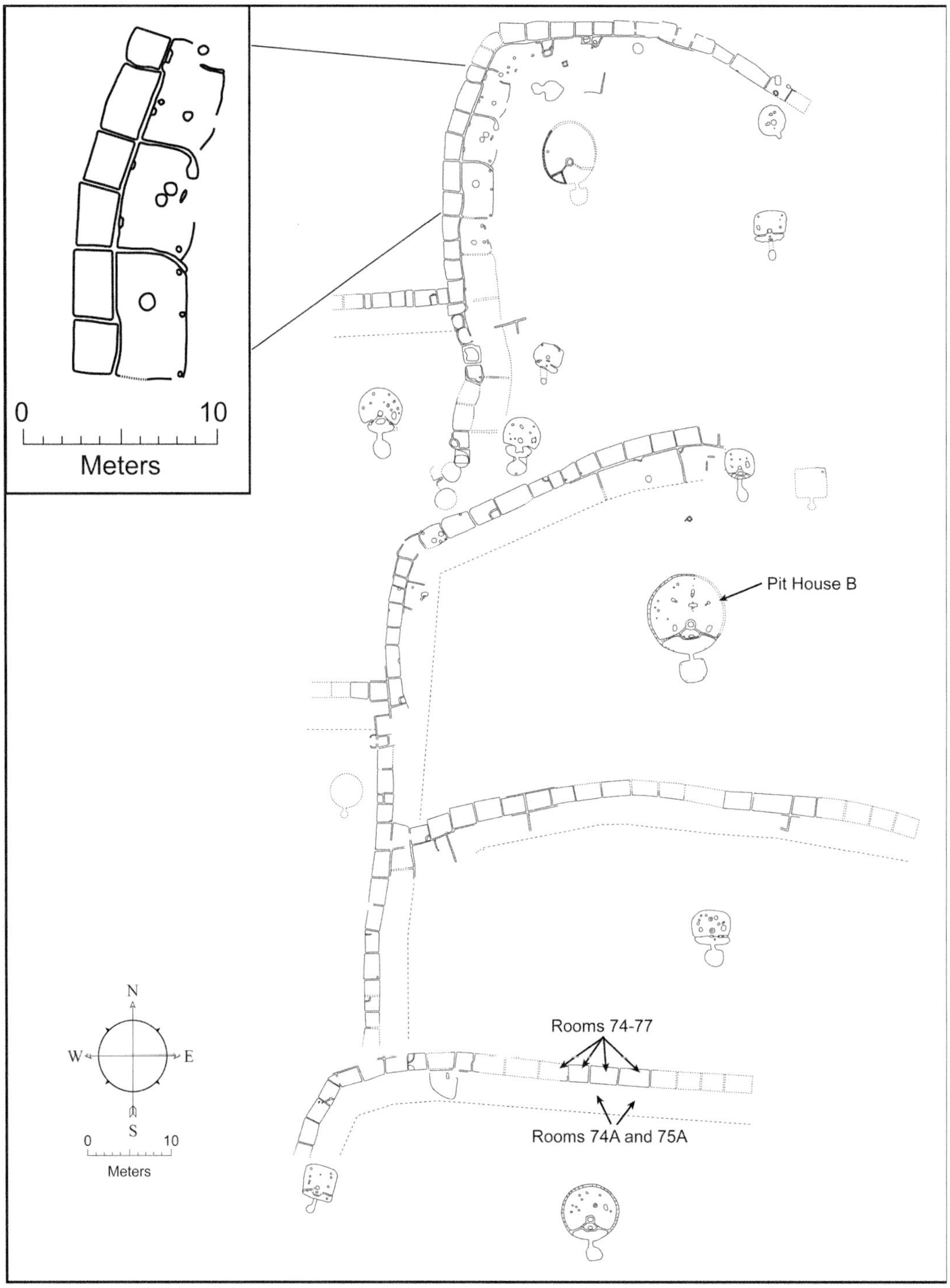


Figure 2. Map of the early Pueblo I village at Alkali Ridge Site 13, as excavated in 1932-1933, based on Brew (1946). The inset at the upper left is an enlargement of a suite of three contiguous houses completely excavated by Brew. The labelled rooms are the locations of our 2012-2013 excavations.

Earthen architecture at Site 13 is vernacular architecture. Much of the site grew by accretion, with little evidence of planning, and almost everyone probably lived in houses they helped build. The largest pit structures, which appear to have been used (at least in part) for communal ritual, can be interpreted as monumental architecture of a sort, but only by comparison to smaller scale domestic architecture on the site. These communal pit structures are much larger than normal houses, and would have required a group effort to build, but even the largest structure at Site 13 is only 60 m2 and would be dwarfed by the monumental earthen architecture discussed by other participants in this volume.

## 2 Technological choices and the built environment

Our goal in this paper is primarily to present a case study, not to delve deeply into theory, but we should acknowledge that our interpretations draw on at least two very different research traditions and bodies of theory. The first of these research traditions is the study of technological systems. Specifically, Lemonnier (1992: 3) emphasizes the importance of studying the technological choices, which he defines as the 'more subtle informational or symbolic aspects of technological systems that involve arbitrary choices of techniques, physical actions, materials, and so forth that are not simply dictated by function...' Related to this is the concept of 'operational sequence' (Lemonnier 1992: 25-27) or '*chaîne opératoire*,' the 'series of operations that transforms a raw material into a product' (Creswell 2010: 26, our translation).

This approach to technological systems originates in twentieth-century French ethnography and archaeology (Mauss 1935; Leroi-Gourhan 1973). It has not been commonly applied in the Southwestern U.S., where we work, although since the 1990s some North American archaeologists have adopted this perspective in the study of lithic and ceramic technologies (e.g., Chazan 2009; Dobres and Hoffman 1994). The second research tradition that influences our approach comes from studies of vernacular architecture by folklorists and architectural historians (e.g., Glassie 1975; 2000; Rapoport 1969). These two research traditions converge in drawing attention to the relationships among architecture, culture, and society. Glassie (2000: 20) declares, for instance, that '...every building is a cultural fact, the consequence of a collision between intentions and conditions...'. Because buildings communicate the builder's intentions and their 'notions of beauty and bodily comfort and social propriety...' (Glassie 2000: 17), architectural form reflects cultural norms and shapes social relations. Leroi-Gourhan makes a similar point when he says that 'the house is at the same time the most visible and the most personal of ethnic traits' (Leroi-Gourhan 1973: 241, our translation).

The concept of technological choices additionally prompts us to look beneath the finished architectural form to understand how the entire operational sequence involved in building is influenced by both culture and circumstances (such as the availability of raw materials or of help from kin or neighbors). This is important in our case study because some buildings at Site 13 would have appeared identical in their finished form, but were built using different materials and techniques. Differences in the operational sequence began with the initial steps of gathering raw materials for house construction, and included differences in the ways walls were built and pit structures were roofed.

## 3 Earthen architecture at Alkali Ridge Site 13

Early villagers at Site 13 made and used four kinds of buildings: storage rooms, surface habitation rooms, residential pit structures, and oversized pit structures. The basic house form consisted of one surface habitation room connected to two or (occasionally) three storage rooms. These houses were above ground, although storage room floors were often slightly below the ancient ground surface. Pit structures were semi-subterranean, with floors well below the ancient ground surface but superstructures that extended above it. They were constructed by first excavating into the sandy sediments that covered the site, and often into the soft sandstone bedrock below. Residential pit structures were relatively small, with between about 10 and 20 m2 of floor area. They may have served as seasonal houses for people who also maintained an above-ground house, or as residences for people who preferred old-style, more traditional house forms (until the mid 700s CE, everyone lived in semi-subterrranean pit houses). They may also have been important for some kinds of ritual, although they were too small to hold many people. The 1932-1933 excavations identified three over-sized pit structures, although there was too little excavation in one of them to estimate its size. The other two were 32 and 62 m2, considerably larger than the residential pit structures, and they have floor features and associations that imply they were used for communal ritual (Allison et al. 2012; Spielmann 2004).

The early Pueblo I village at Site 13 consists of four arcing double rows of surface rooms, with pit houses in the open plaza spaces that were defined by the surface rooms (see fig. 2). Brew (1946: 189) described the surface roomblocks as 'long rows of contiguous small rooms, roughly rectangular, arranged in a gentle curve until the west end was reached, where the rows either curved sharply to the south or abutted a north-south running row of similar rooms'. Surface habitation rooms are arranged in rows along the insides of the arcs. They are larger than the adjoining storage rooms, which create the outside row of the arcs. The 1932-1933 excavations, and our own, focused most of their attention on the storage rooms and so we will discuss them first and use them as a framework to discuss the habitation rooms.

### *3.1 Storage Rooms*

The most obvious variation in construction techniques is in storage room walls. This may be because far more of them have been excavated than any other building form. They are interpreted as storage rooms because they are smaller than the habitation rooms, lack hearths, and apparently could only be entered through an adjoining habitation room, which always had hearths. The excavated storage rooms also contained large amounts of maize and beans, and occasionally pinyon nuts and other seeds (Brew 1946: 190), making it clear that they were, in fact, used for storage. Many complete or restorable ceramic vessels were also recovered from these rooms.

The act of constructing a storage room began with accumulating materials and preparing the ground surface. Storage room floors were, as Brew (1946: 190-191) explains, 'as a rule 8 to 15 centimeters below the original ground level'. This initial excavation would have had the benefit of providing earth for the later stages of construction. The floors were left earthen during the use-life of the structures.

Brew (1946: 191-193) identified five varieties of wall architecture in the excavated storage rooms (Figure 3). The most common type, which he labeled Type I, was constructed with a single course of upright slabs at the base, lining and extending above, the shallow rectangular pit in which the room was built. Above the slabs, alternating courses of thin stones and what Brew called 'turtleback adobe' formed the upper portions of the walls. There is, however, no evidence that adobe bricks were prepared and dried ahead of time for use in these walls. Instead, still-wet lumps of clay mixed with water were probably placed on the wall.

The much less common Type II walls are similar to Type I, but with two courses of smaller slabs at the bottom. Type III walls have alternating earth and thin stones, like the upper portions of Type I and II walls, but with no slab-lined base. Type IV walls have closely placed wooden posts plastered with earth. The few Type V walls are made solely of 'adobe'; but again, these walls were apparently formed from wet-laid lumps of clay and are therefore cob rather than adobe. When finished, all of these walls were covered with a layer of clayey plaster up to 10 cm thick. It was not possible to tell which method of construction was used on a finished storage room wall just by looking at it. Only those who witnessed the construction, or participated in it, would have known which method was used.

The operational sequences involved in building each of these wall types were different, although Types I, II, and III were similar. Differences began with assembling raw materials. Every wall type required an abundant supply of clayey dirt and water to mix it with. In the semi-arid Southwestern U.S., this probably would have required hauling water from nearby springs. The closest one is about 700 m west of the site. In an experimental replication of early Pueblo earthen surface structures similar to the ones at Site 13, Varien (1984: 39-41) found that collecting and transporting water was more costly than the combined cost of assembling all the other materials needed. The cost of obtaining water could have been reduced by using rainwater collected on site, or by using earth that was already wet from melting snow or recent heavy rains. Those options would have required

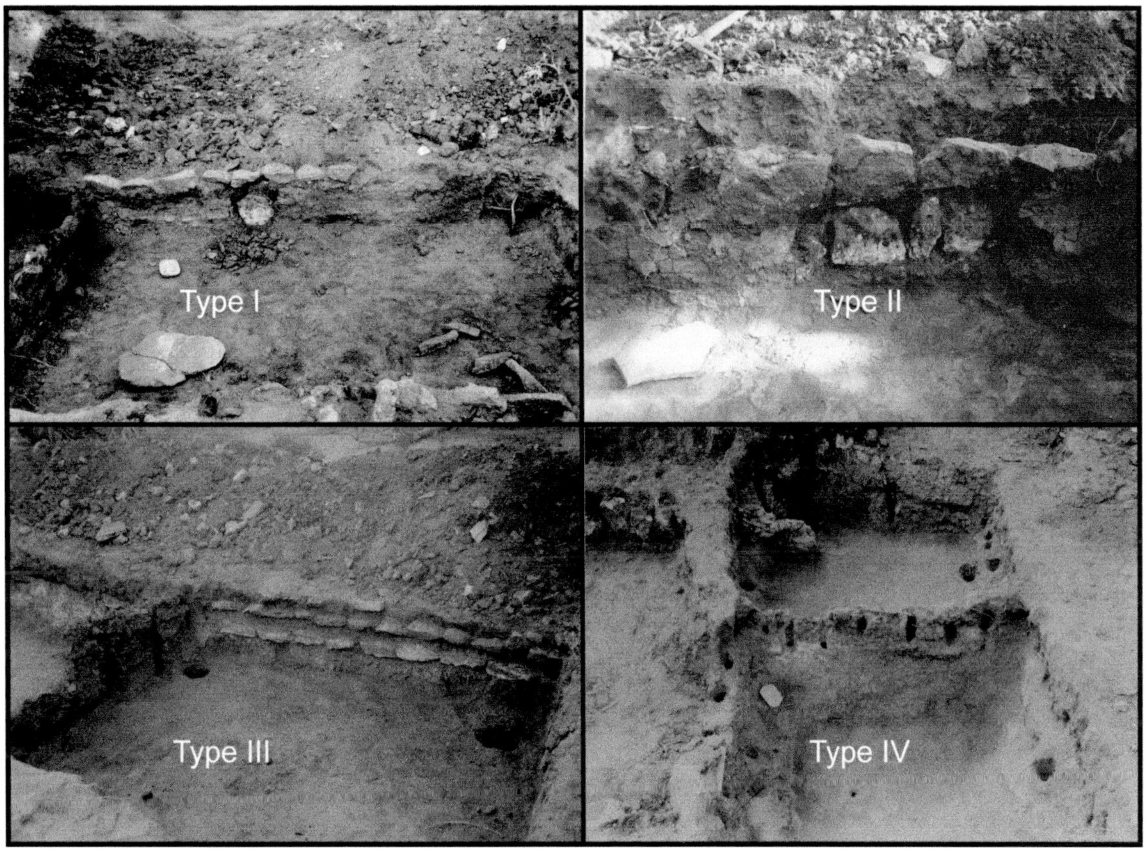


Figure 3. Photographs of four of the different storage room wall types encountered in the 1932-1933 excavations. The bottom of the Type I wall, with a single row of slabs at the base, is still covered with a thick payer of clay plaster, except where the excavators dug through it (just left of the center of the photograph) to reveal a slab at the base. Note the double row of slabs at the base of the Type II wall, the lack of basal slabs in the Type III wall, and the closely spaced post holes in the Type IV wall (courtesy of the Peabody Museum of Archaeology and Ethnology, Harvard University, PM 33-7-10/1380.1.152, PM 33-7-10/1380.1.153, PM 33-7-10/1380.1.169, and PM 33-7-10/1380.1.226).

waiting for the right conditions and would have restricted building to the early spring when the ground held moisture, or to the late summer when thunderstorms are common.

Wood for roof beams and roof support posts would also have been required for every wall type. Storage room roofs were partially supported by their walls, but every storage room also has four small posts located near each of the four corners of the room. The postholes average 13 cm in diameter. To build a storage room with Type IV walls would have required amassing considerably more wood to make the small posts that were embedded in the wall.

Our excavations in 2012 and 2013 included reexcavating three storage rooms that had originally been excavated in 1932, as well as a fourth one that had not been previously excavated. All of the rooms had burned, and we recovered abundant pieces of burned wood. It was difficult to specifically separate roof beams from posts, except in the case of two post remnants that were still in their postholes. These were both of juniper, as were about half of the recovered pieces of wood. The other half of the recovered structural wood was either pinyon pine or ponderosa pine, which occurred in approximately equal amounts. Today, and probably in the past, juniper and pinyon grow abundantly near the site, while ponderosa pine grows at higher elevations, and would have needed to be carried over longer distances of at least several kilometers.

Building a storage room with Type I, II, or III walls would also have required a supply of appropriately sized stones and sandstone slabs. Despite the use of stone, these walls are not masonry; the greater portion of the walls were earthen and the stone did not support the weight of the building. The

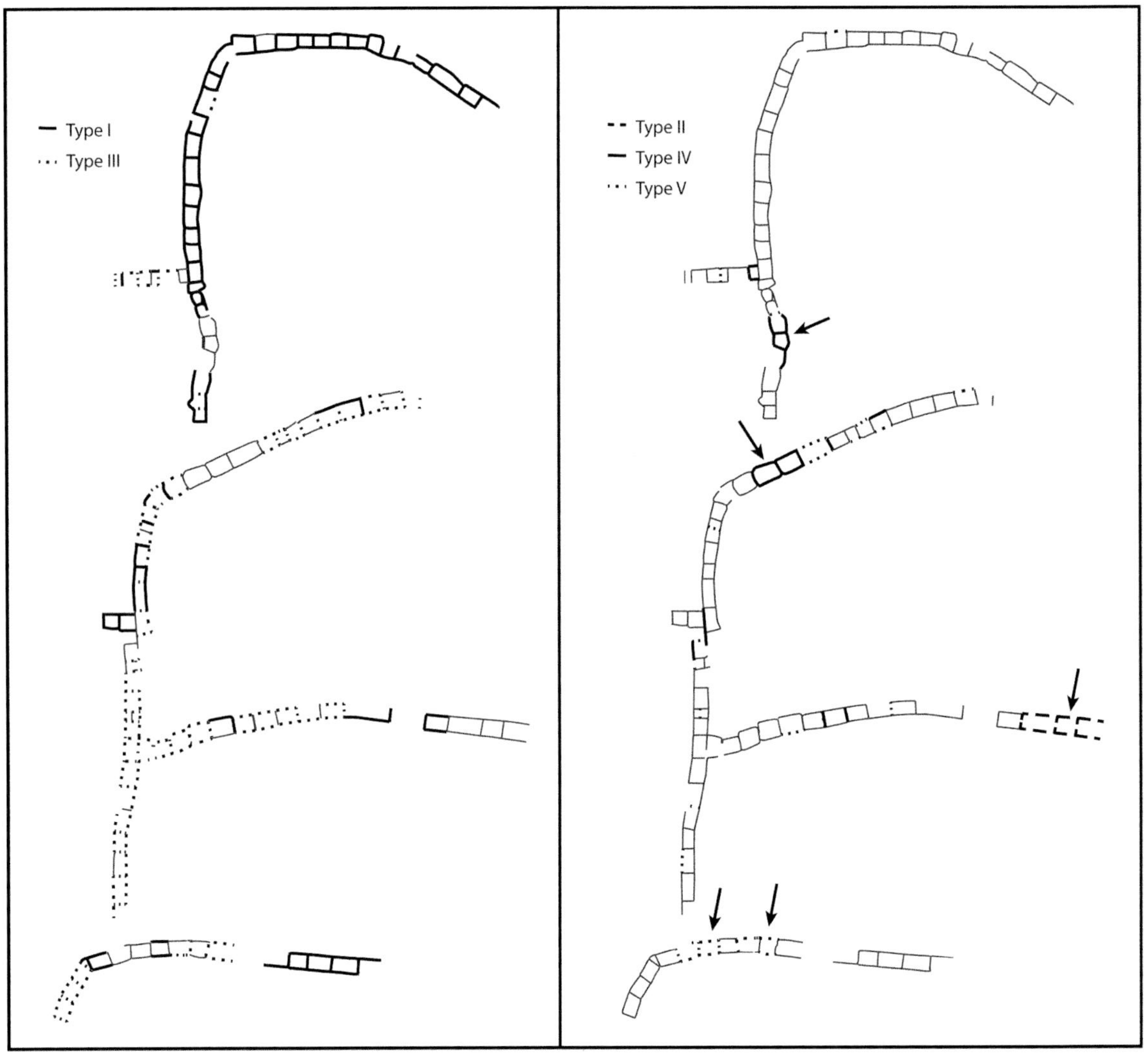


Figure 4. Schematic maps of the early Pueblo I village at Site 13 showing the distributions of the different storage room wall types. On the left, the two most common wall types, which have very different distributions in the site (in particular, note the difference between the northern roomblock, with almost all Type I walls, and the remainder of the site). On the right, the three less common wall types; the arrows point to what appear to be houses built with Type II, Type IV, or Type V walls.

slabs at the base of Type I and II walls probably were intended to (and probably did) impede access by rodents or other pests to foods kept in the storage rooms, but they were only about 5 cm thick, too narrow to build the upper part of the wall upon. Instead, thick clay plaster over the slab both hid it from view and provided a base wide enough to support the upper wall. The smaller stones that were laid in rows in the upper wall seem to have been chinking. They would have reduced the amount of clay and (probably more important) the amount of water needed to build a wall, and perhaps helped control shrinkage and cracking as the wet-laid cob dried.

The spatial distributions of the different wall types exhibit several interesting patterns across the site (Figure 4). At a large scale, Types I and III, the most common wall types, have complementary distributions within the site. The slab-lined Type I walls make up 79% of the walls in the northern arc, but only 17% of the walls in the central section of the site. Type III walls, of coursed masonry and 'adobe', comprise only 10% of the walls in the northern arc, but 58% of the walls in the central section. The remaining three wall types are scattered throughout the site.

In a few places, details of the walls indicate the building sequence. Our 2012 and 2013 fieldwork showed that the long back and front walls of the four storage rooms we excavated were built at

Figure 5. Photograph of Rooms 74-77, facing east (Room 74 is at the top, Room 77 at the bottom). Rooms 74-76 were excavated in 1932; we reexcavated them in 2012. Room 77 was not excavated in the earlier excavations, and is only partially excavated in this photograph. Note that the rows of slabs in the long northern and southern walls of these rooms (to the left and right in the photograph) are continuous, and the slabs in the cross walls abut them. All of these slabs were covered with plaster when first excavated, but 80 years of post-excavation erosion has left the slabs bare. The Type I wall in Figure 3 is the northern wall of Room 75 (in the photograph, the wall on the left of the second room from the top) of post-excavation erosion has left the slabs bare. The Type I wall in Figure 3 is the northern wall of Room 75 (in this photograph, the wall on the left of the second room from the top).

once, and then cross walls were added to divide the space into rooms (Figure 5). The way these storage rooms correspond with adjacent habitation rooms indicates they were part of three different houses. So in this part of the site, at least, people appear to have worked together to build several houses at the same time. On the other hand, pairs of storage rooms built from the less common wall types show that much of the site did grow piecemeal through (presumably unplanned) addition of individual houses (see Figure 4).

Our excavation recovered much burned clay construction material that was formerly part of the walls and roofs. This earthen material is the clay-rich architectural mud that made up the majority of the walls and was part of the roof. Some of it carries the imprint of beams and other perishable materials. It filled in the cracks of the wooden structures and provided a measure of insulation and protection from the weather. The fire that destroyed the site hardened the clay into a low-grade ceramic, preserving it and the cavities created by the organic inclusions or the material that made up the framework of the roof. Over 350 pieces were recovered from two habitation rooms, three storage rooms, and a pit structure. The analysis of this material consisted of a number of measurements of the fired clay and its impressions (Figure 6). Identifications of the material that created the impressions were made whenever possible. An analysis of this material allows us to make some determinations about the walls above the foundations, particularly the framework that has since been lost to destruction and time.

We were able to collect impressed clay construction material from three of the four storage rooms we excavated in 2012 and 2013 (Rooms 74, 75, and 77 in Brew's numbering system). Each

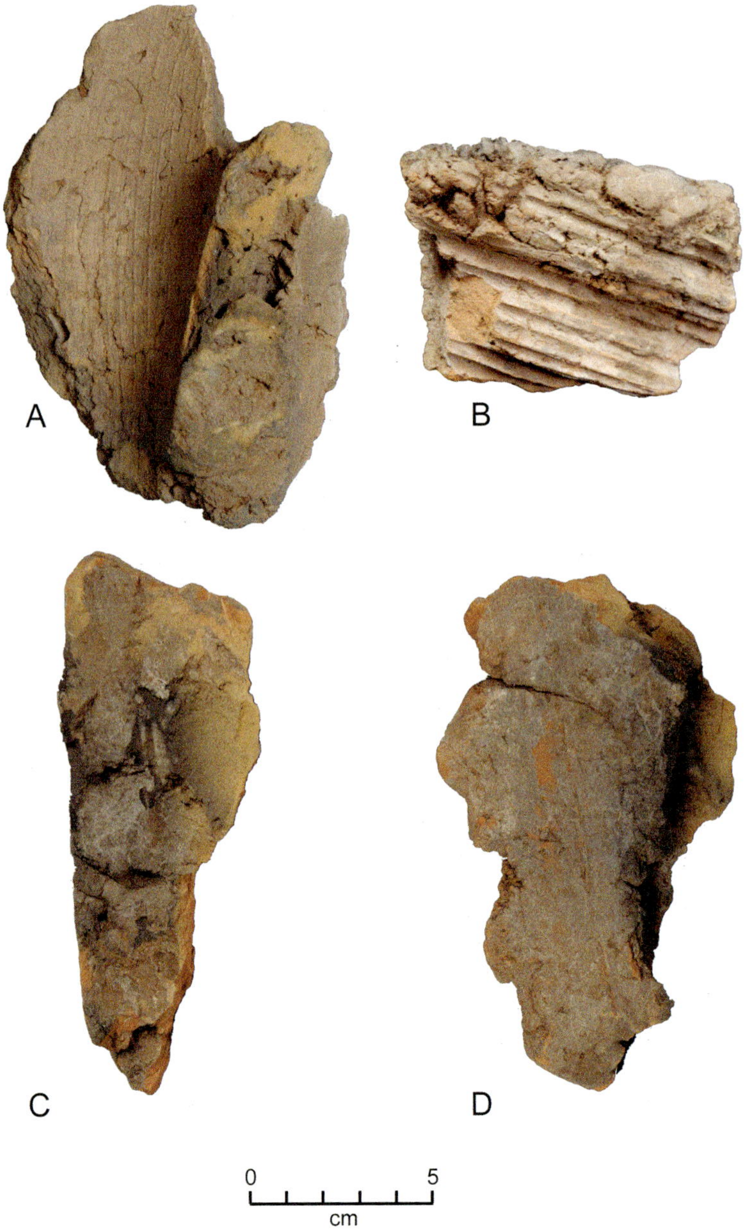


Figure 6. Fire-hardened daub from storage rooms (probably mostly from the roofs of those rooms), providing clues about construction techniques and materials. A is from Room 74 and shows a large concave impression of an unhusked ear of maize. A smaller impression of a maize cob is visible toward the right side of the piece just above the center. B is from Room 77 and shows the impression of what appears to have been a bundle of sticks. C and D are two views of the same piece from Room 77. On the left, are impressions left by the builder's fingers. On the right, a flat impression shows grain that suggests a wooden board, probably made by splitting a log.

of these storage rooms was part of a separate house and connected to a different habitation room, although, as noted above, the lower walls (at least) of these rooms were constructed together. The impressed clay from the storage rooms seems to come mostly from the roofs, which were framed with roof beams that rested on top of the walls, with additional support from the corner posts. Clay was pressed between and around the beams, and the beams were covered by smaller materials (such as reeds and sticks), which were in turn covered by more clay.

The fire-hardened clay construction material from Storage Room 74 has an average thickness of 2.1 cm and beam impressions with diameters ranging from 4-7 cm. The material from the roof of Room 75 has an average thickness of 1.1 cm, with impressions of beams that ranged from 3-6 cm in diameter. Burned daub from the roof of Storage Room 77 has an average thickness of 2.4 cm and beam diameters ranging from 6.5-14 cm. The average diameters of beam impressions from the three rooms were consistently small, at 4.7, 4.5, and 4.5 cm, respectively. The impressions indicate that a few larger beams were present (up to 14 cm in diameter), but the roof beams of these small rooms did not need to be large, and most were less than 5 cm in diameter.

There are also impressions in the daub that are not strictly architectural, but can inform us about construction techniques. Marks left by fingers in clay indicate that mud was applied to the wooden frame of the roof by the handful, pressed on, and smoothed on the exterior (see fig. 6C). The impressions of fingers on clay from our excavation do not appear to be decorative in nature, although Brew found that many of the rooms had interiors plastered with a layer of clay and decorated, apparently to personal taste, with incised lines, rolled corncob impressions, or series of dimples from pressed fingers.

Additionally, it appears that a small amount of vegetal material was incorporated into the matrix of the clay construction material, possibly as an intentional organic temper. This includes reeds, sticks, grasses, leaves, as well as corn cobs and corn husks (see Figure 6a and 6b). There are even some instances of charred corn kernels encased in the matrix. Most of these elements were fully encased in the clay rather than having been pressed into the surface.

The presence of organic temper can provide tensile strength, distribute tension during the drying process, and provide a path for moisture so that the clay dries more evenly (Allerbo and Waldemarsson 2013: 11; Binici, Aksogan and Shah 2005: 313; Cooke 2010:13). Adding as little as 4% fiber into the mix prevents visible cracks from appearing and adds strength to the structure (Ghavami, Toledo and Barbosa 1999: 39). The amount of vegetal material in the daub from Site 13 may have fallen around or below the 4% mark. The low level does not fully utilize the benefits of organic temper, but it does help avoid the inevitable weakness that is caused by the deterioration of the vegetal material (Cooke 2010: 33).

### *3.2 Habitation Rooms*

Brew excavated 25 surface habitation rooms in 1932 and 1933, although he paid more attention to storage rooms because of their easily defined walls and abundance of artifacts; many of the surface habitation rooms were left unexcavated, even where the storage rooms from the same houses were dug. We excavated portions of two other surface habitation rooms in 2012 and 2013, adjacent to the four storage rooms we cleared. The habitation rooms are identified by their relatively large size, and the presence of hearths. They are located on the interior of the roomblock curve. Unlike the slightly subterranean storage rooms, the floors of habitation rooms are at the ancient ground surface. The higher floors made the walls and floors more susceptible to erosion, consequently making these rooms harder to identify and define.

These living areas, like the storage rooms, appear to have been made with a variety of construction techniques. Interestingly, habitation room wall styles are often different from those used to build the connected storage rooms. For example, the northern two roomblocks have habitation room walls that are almost exclusively Type IV, with earthen material between and over a framework of posts, despite the storage room walls in that section being predominately Types I and III. Few habitation rooms were excavated in the southern two rows, but wall types include Types I, IV, V, and an idiosyncratic wall that has two worn-out metates among the upright slabs that make up the base of the wall (Brew 1946: 195-196).

Our excavations in 2012 and 2013 explored small sections of two rooms, which we have numbered 74A and 75A (following Brew's convention of designating habitation rooms with an A appended to the number of one of the connected storage rooms). The wall between the two habitation rooms is a Type III wall, having rows of small stones with layers of earth in between. The wall starts at, or near, the prehistoric ground level.

More fire-hardened architectural materials were recovered from the habitation rooms than the storage rooms despite the habitation rooms being only partially excavated. Material from the roof of Room 74A was an average of 2.6 cm thick with beam diameters ranging from 8-14 cm. Roof material from Room 75A averaged 3.1 cm thick, with beam impressions ranging from 5-16 cm. The average diameter of the beam impressions was 10.1 cm for Room 74A and 9.2 cm for 75A. In comparison to their connected storage rooms, the habitation rooms have thicker roofs, both in terms of a thicker layer of clay on top as well as beam widths that are about twice the diameter as the storage room roof beams.

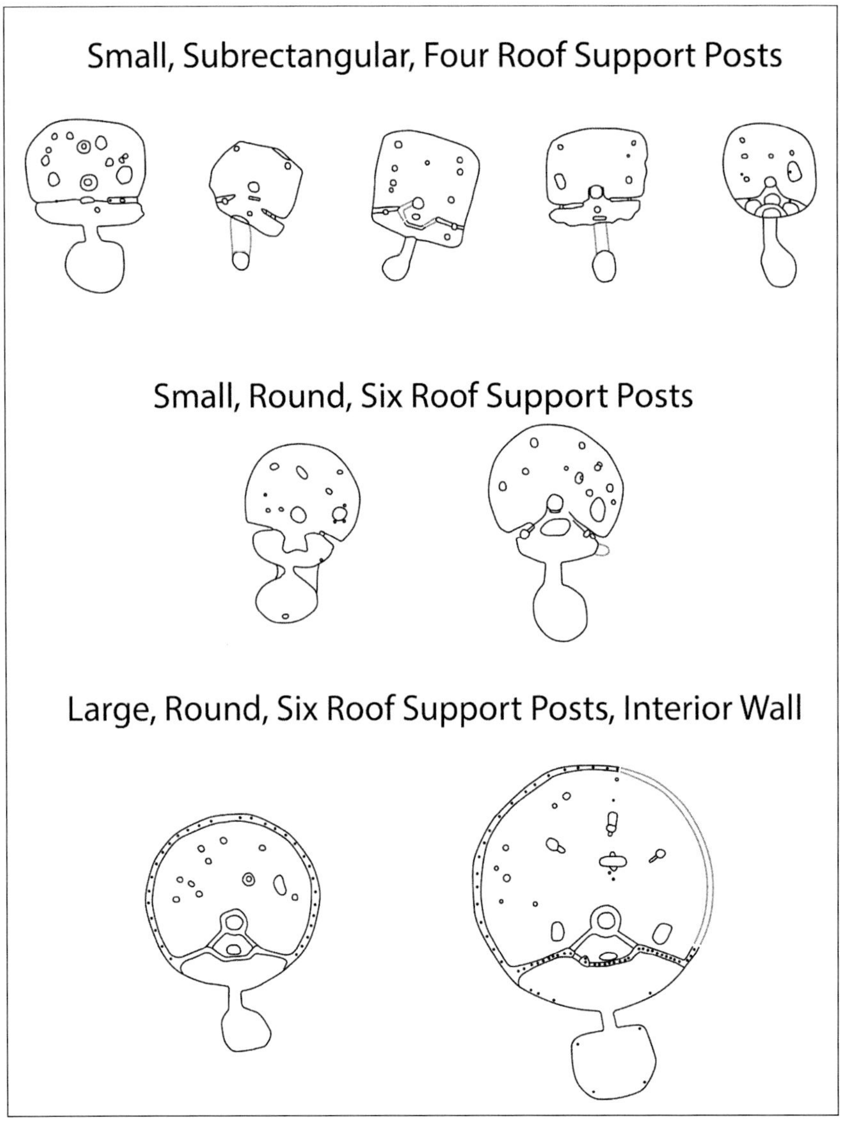


Figure 7. Variation in size, shape, and roof design among early Pueblo I pit structures at Site 13.

### *3.3 Residential pit structures*

The 1932-1933 excavations at Site 13 identified and at least partially excavated eight small pit structures that appear to be residences of some kind. These structures are generally located in the courtyard areas in front of the curved roomblocks. Our own recent excavations did not include any residential pit structures. The smallest of the eight excavated small pit structures is only 8 m2, irregular in shape, and has no obvious roof support posts. It may have been a temporary shelter roofed flat at ground level. The other seven range in size from 10 to 21 m2.

While we cannot say much about them, these apparently residential pit houses vary in shape and in the way their roofs were constructed (Figure 7). Three of them are round, three are subrectangular, and one is D-shaped. It seems significant that the round pit structures all had six main roof supports set into the floor, while the subrectangular and D-shaped ones all had only four roof supports. It is not clear whether these differences in structure shape and roof design would have been visible from outside the structure.

### *3.4 Oversized pit structures*

Three pit structures excavated in 1933 were distinguished by their size, as well as details of their design. One of these (Pit House M) was not thoroughly excavated and it is impossible to estimate its size, although the curvature of the wall suggests it was large. Pit House E was 32 m2, half again as large as the largest of the residential pit structures, while Pit House B was almost twice as big as E, at 62 m2. All of the oversized pit structures are round, and B and E have (at least) six roof supports. Unlike the smaller, apparently residential pit structures, all three of the oversized pit structures have a post-and-daub wall lining the inside of the pit. The oversized pit structures also have floor features with apparent symbolic meaning that indicate ritual use of the structures (Wilshusen 1988; 1989). Most of Pit House B, the largest structure, was excavated in 1933, but a baulk of unexcavated fill was left in the northeastern part of the structure. In 2013 we dug a 1 m wide trench through this unexcavated fill, and we were able to observe many architectural details.

The original pit for Pit House B was excavated down to (and probably slightly into) the soft sandstone bedrock underlying the site. A layer of plaster laid over the stone formed the floor and subfloor features were cut into the stone, including the post holes for the six main roof supports. A series of floor-level post holes encircle the main chamber and indicate the presence of a post-and-clay wall lining the edge of the structure. Our excavation in Pit House B recovered a great deal of fire-hardened roofing material, as well as pieces that apparently came from the wall that encircled the structure. Both the number of pieces and the size of the specimens were impressive. Finger impressions show that this material was applied to the roofing beams by hand. The clay material from the roof was 4.1 cm thick, supported by beams with diameters ranging from 5-27 cm with an average of 11.8 cm.

Measurements from hardened clay show that daub from the roof of the Pit House B is twice as thick as from the roofs of the above-ground structures. Larger beams were also used in the construction of Pit House B than the surface habitation structures (9.2 and 10.1 cm average for the surface habitation structures and 11.8 cm for Pit House B; fig. 8a and 8b). 18% of Pit House B's beams (N=29) are larger than 15 cm while only one beam from any of the surface structures is that large. The larger roof beams added strength to span the wider distance and to make up for the weight of the thicker clay.

One large fragment from the encircling wall (Figure 8c and 8d) is particularly informative. It is about 7 cm thick and 18.5 cm wide, with post impressions on both sides showing it was pressed between the posts against the wall of the excavated pit. Finger impressions show where the clay was pushed out against and over the post on one side. The posts in the encircling wall were apparently about 10 cm in diameter, and were spaced about 20 cm apart; where our 1 m wide trench encountered the structure wall, it uncovered four post holes (Figure 9).

## 4 Discussion and conclusion

We have described, in some detail, the vernacular architecture of one early Pueblo village, Alkali Ridge Site 13. Despite the relatively simple nature of construction technology at the site, there is considerable variation in how people built their buildings. Variation among structure types relates to a combination of functional considerations and ideas about how different kinds of structures should be built. Some of the variation in construction techniques is no doubt functional. Lining the bottom of storage rooms with sandstone slabs, for instance, is easily explicable as a technique for keeping pests out. But why not do it for every storage room? In this, and in other characteristics, structures at Site 13 vary in ways not easily explicable in terms of function. In particular, variation in storage room wall types and pit structure shape and roofing seems to represent technological choices by builders selecting among functionally equivalent alternatives.

If (as we quoted earlier) 'every building is a cultural fact, the consequence of a collision between intentions and conditions' (Glassie 2000), then understanding the variation in building styles at Site 13 should take account of builders intentions (to the extent possible) and the conditions in which they created buildings. One important condition is the social context of Site 13. Prior to the 750s CE, when the site began to be built, settlements were small and more dispersed. The

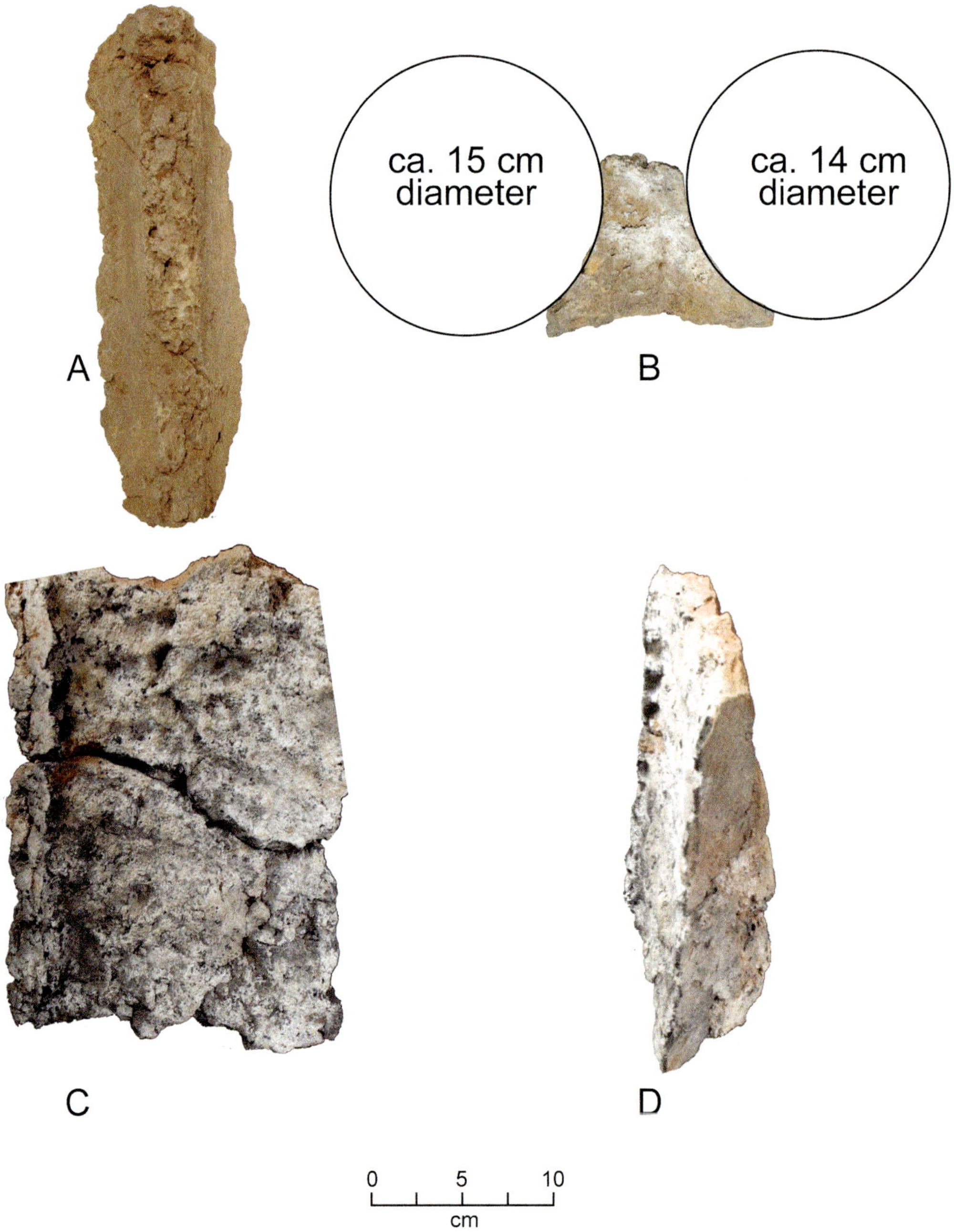


Figure 8. Fire-hardened daub from Pit House B. A and B are the same piece, probably from the roof. The piece is flat on one side and had been pressed between two large beams, which left large enough impressions to allow estimation of their diameters. C and D are two views of the largest piece of daub, which clearly came from the wall that lined the inside of the pit, encircling the structure. The left and right edges in C show impressions from posts, one of which is visible on the right in D. This piece also has numerous impressions from a builder's fingers (by the left edge in both photos, where the plaster was pressed against, and at least partially over, a post).

aggregation of several hundred people into a village of contiguous houses such as Site 13 involved considerable change in the ways people interacted with each other. In particular, for the first time, groups of people too large to all be closely related lived together in one settlement. It also follows a period in the early 700s CE when regional population is low.

Populations increase rapidly and the earliest villages form in several parts of the Mesa Verde region in the late 700s CE, at about the same time as people moved to Site 13. In the eastern part of the Mesa Verde region, 130 km southeast of Site 13, Potter and Yoder (2008: 29) describe variation

Figure 9. East end of our 2013 trench in Pit House B, showing the intersection of the structure floor and the encircling wall. The arrows point to four post holes (only one of which is excavated) that were present in this 1 m-wide section of the wall.

in pit structure construction methods at early Pueblo I sites in Ridges Basin (dating to almost exactly the same time period as Site 13). They interpret this variation as 'the result of immigrant households from various origins actively working to establish and signal their identities.'

Architectural variation at Site 13 suggests that the village was built and lived in by people with different ideas about the proper way to build a storage room, the best shape for a pit structure, or how many support posts a pit structure needs. This reflects the fact that the village formed through the aggregation of numerous small social groups with diverse origins and local traditions. But it seems important that much of this variation was invisible in daily life. People could not use differences in storage room wall techniques to actively 'establish and signal their identities'—the differences were plastered over, and the storage rooms were the back rooms of houses that would only be visible to household members or guests. Similarly, it is not at all clear that the internal shape of a pit structure or the number of roof support posts it had would be visible without entering the house.

The technological choices evidenced in Site 13 architecture reveal diversity in builders' intentions, and at one level the observed differences probably reflect builders' social identities, grounded in diverse origins. But the outward appearance would be homogenous, not diverse. It signals the downplaying of the diversity in origin of the villagers and the active building of a community-level social identity based on co-residence.

## Acknowledgements

Many colleagues made important contributions to the 2012-2013 excavations at Site 13 and the subsequent analyses. In addition to the authors, the excavation staff included Katie Richards, Lindsay Johansson, Megan Mangum, and Stephen Freeman, all of whom were then graduate students at Brigham Young University (BYU). We were aided in the field by volunteers too numerous to list here, most of whom were professional or avocational archaeologists from the local area. Several volunteers at the BYU Museum of Peoples and Cultures completed the analysis of burned daub, including Brandon Walter, Lori Hart, and Kevin Holm. Marion Forest completed the digitization of the paper analysis sheets. Scott Ure drafted the site map based on the original from Brew (1946) that was the basis for figures 2 and 4. He also drafted the structure maps in Figure 7. The other figures, unless otherwise stated, are ours. The 2012-2013 excavations were authorized by a permit issued by the United States Bureau of Land Management (ARPA permit 12UT85077), and BLM archaeologist Don Simonis provided logistical support and encouragement. Funding for the excavation was provided by grants from the BYU Department of Anthropology's Shallit fund. Mark Varien read the first draft of this paper and offered valuable comments. Finally, we want to thank the session organizers, Annick Daneels and Maria Torras Freixa, for including us and for their work editing and formatting our original, not-quite-ready, submission.

## Bibliography


Allerbo, E. and Waldemarsson, H. (2013) Adobe-Technology of Everybody. Unpublished bachelor's thesis. Halmstad University.

Allison, J.R. (n.d.) 'Dating Early Pueblo I Villages in Southeastern Utah: Insights from Collections, Archives, and Fieldwork' in Nash, S.E. (eds) Pushing Boundaries: Proceedings of the 2018 Southwest Symposium (under review).

Allison, J.R., Hurst, W.B., Till, J.D. and Irwin, D.C. (2012) 'Meanwhile, in the West: Early Pueblo Communities in Southeastern Utah' in Wilshusen, R.H., Schachner, G. and Allison, J.R. (eds) *Crucible of Pueblos: The Early Pueblo Period in the Northern Southwest.* Los Angeles: Monograph 71. Cotsen Institute of Archaeology Press, University of California: 35-52.

Binici, H., Aksogan, O. and Shah, T. (2005) 'Investigation of Fibre Reinforced Mud Brick as a Building Material', *Construction and Building Materials*, 19(4): 313–318.

Brew, J.O. (1946) *Archaeology of Alkali Ridge, Southeastern Utah.* Papers of the Peabody Museum of American Archaeology Vol. 21. Cambridge: Harvard University.

Chazan, M. (2009) 'Pattern and Technology: Why the Chaîne Opératoire Matters' in Shea, J.J. and Lieberman, D.E. (eds) *Transitions in Prehistory: Essays in Honor of Ofer Bar-Yosef.* Oxford: Oxbow Books: 467-476.

Cooke, L. (2010) *Conservation Approaches to Earthen Architecture in Archaeological Contexts.* Oxford: BAR International Series 2147.

Creswell, R. (2010) 'Techniques et culture', *Techniques et culture* 54-55(1): 23-45. [accessed July 25 2019 at http://journals.openedition.org/tc/4979; DOI: 10.4000/tc.4979]

Dobres, M.A., and Hoffman, C.R. (1994) 'Social Agency and the Dynamics of Prehistoric Technology', *Journal of Archaeological Method and Theory*, 1(3): 211-258.

Ghavami, K., Toledo Filho, R.D. and Barbosa, N.P. (1999) 'Behaviour of Composite Soil Reinforced with Natural Fibres', *Cement and Concrete Composites*, 21(1): 39-48.

Glassie, H. (1975) *Folk Housing in Middle Virginia.* Knoxville: University of Tennessee Press.

Glassie, H. (2000) *Vernacular Architecture.* Bloomington and Indianapolis: Indiana University Press.

Lemonnier, P. (1992). *Elements for an Anthropology of Technology.* Ann Arbor: The University of Michigan Press.

Leroi-Gourhan, A. (1973) *Milieu et Techniques.* Paris: Éditions Albin Michel.

Mauss, M. (1935) 'Les Techniques du Corps', *Journal de Psychologie Normale et Pathologique*, 32: 271-293.

Potter, J.M. and Yoder, T.D. (2008) 'Space, Houses, and Bodies: Identity Construction and Destruction in an Early Pueblo Community' in Varien, M.D. and Potter, J.M. (eds) *The Social Construction of Communities: Agency, Structure, and Identity in the Prehispanic Southwest.* Lanham: AltaMira Press: 21-40.

Rapoport, A. (1969) *House Form and Culture.* Englewood Cliffs: Prentice-Hall.

Spielmann, K.A. (2004) 'Communal Feasting, Ceramics, and Exchange' in Mills, B.J. (ed.) *Identity, Feasting, and the Archaeology of the Greater Southwest.* Boulder: University Press of Colorado: 210-232.

Varien, M.D. (1984) Honky House: The Replication of Three Anasazi Surface Structures. Unpublished Master's thesis, University of Texas.

Wilshusen, R.H. (1988) 'Sipapus, Ceremonial Vaults, and Foot Drums (or a Resounding Argument for Protokivas)' in Blinman, E., Phagan, C.J. and Wilshusen, R.H. (compiled) *Dolores Archaeological Program: Supporting Studies: Additive and Reductive Technologies.* Denver: United States Department of the Interior, Bureau of Reclamation: 649-671.

Wilshusen, R.H. (1989) 'Unstuffing the Estufa: Ritual Floor Features in Anasazi Pit Structures and Pueblo Kivas' in Lipe, W.D. and Hegmon M. (eds) *The Architecture of Social Integration in Prehistoric Pueblos.* Occasional Papers No. 1. Cortez: Crow Canyon Archaeological Center: 89-111.

Wilshusen, R.H. (2018) 'How Agriculture Took Hold in the Mesa Verde Region: A Review of Recent Research on the Late Basketmaker-Early Pueblo Periods (A.D. 500-920)', *Reviews in Colorado Archaeology*, 1: 69-95.

Wilshusen, R.H., Hurst, W. and Chuipka, J. (2012) 'Small Hamlets to Early Great Houses: The Emergence of Villages in the Mesa Verde Region between AD 200 and 900' in Young, L.C. and Herr, S.A. (eds) *Southwestern Pithouse Communities, AD 200-900.* Tucson: University of Arizona Press: 141-154.

Wilshusen, R. H., and Potter, J.M. (2010) 'The Emergence of Early Villages in the American Southwest: Cultural Issues and Historical Perspectives' in Bandy, M.S. and Fox, J.R. (eds) *Becoming Villagers, Comparing Early Village Societies.* Tucson: University of Arizona Press: 165-183.

## Chapter 9

# Earthen mounds and political centres: the rise and fall of the Izapa Kingdom, Chiapas, Mexico

Robert M. Rosenswig
University at Albany - SUNY
rrosenswig@albany.edu

**Abstract**

Mesoamerica is one of the cradles of civilisation where kingdoms and states emerged during the latter part of the first millennium BCE. Recently collected lidar (light detection and ranging) and archaeological survey data document for the first time the entire kingdom of Izapa, on Mexico's southern Pacific coast. New settlement survey data track changing distribution of population from 1000 BCE-100 CE and indicated the city's population reached 5725 inhabitants. Izapa was the capital of a regional kingdom with more than 40 lower-order monumental centres that were built according to the same design principles and together formed an administrative hierarchy within the kingdom. Recent dating confirms the apogee of the kingdom at 300-100 BCE and volcanological reconstruction suggests that the Tacaná volcano eruption corresponds with archaeological evidence of political and demographic disruptions to the kingdom. This study exemplifies the transformative capability of lidar technology for documenting earthen mound in tropical environments.

**Keywords:** Pre-Columbian Archaeology, State formation, Urbanism

**Résumé**

La Mésomérique est un des berceaux de la civilisation, où les royaumes et les états émergent au cours du premier millénaire avant notre ère De récentes données de reconnaissance terrestre et áerienne par lidar (light detection and ranging) ont permis pour la première fois de définir l'extension du royaume d'Izapa, sur la côte Pacifique du Mexique. Ces données permettent de suivre le développement de la population entre 1000 AEC et 100 AEC, atteignant jusqu'à 5725 habitants. Izapa fut la capitale d'un royaume gouvernant plus de 40 sites subordonnés, dont le tracé des constructions monumentales réplique les mêmes principes architecturaux, ce qui reflète l'existence d'une hiérarchie administrative. Les études chronométriques récentes confirment l'apogée du royaume entre 300 et 100 AEC, et permettent de faire le lien entre les altérations démographiques et politiques observées et la sequence d'éruptions du volcan Tacaná. Cette étude démontre la particulière utilité de la technologie lidar pour enregistrer des sites composés de monticules artificiels de terre dans une végétation tropicale.

**Mots-clés:** archéologie précolombienne, formation de l'état, urbanisme

## 1 Introduction

Mesoamerica is one of a handful of locations in the world where urban life was independently established (Cowgill 2004). The first research projects to contextualise early Mesoamerican cities with regional settlement patterns were in the Mexican highlands (Sanders, Parsons and Santley 1979; Kowalewski et al. 1989). Due to relatively dry conditions and extensive erosion, evidence of Pre-Columbian occupation is visible on the ground surface in these areas. In contrast, the settlement around most of Mesoamerica's lowland states is much harder to detect, hidden in remote rain forest. Lidar (light detection and ranging) allows for unprecedented accuracy and efficiency in collecting regional settlement data, especially in environments covered in thick, tropical vegetation (e.g., Chase et al. 2012; Evans et al. 2013). Here we report the results of nearly 600 sq km of lidar that were employed in conjunction with pedestrian survey to document the kingdom of Izapa in the Soconusco region of southern Mexico (Figure 1).

Izapa was the largest of a string of urban kingdoms that coalesced on the Pacific coast of Mesoamerica during the first millennium BCE (Clark and Pye 2000; Love 2007). Izapa has long been known to have dominated the Soconusco region on the Pacific coast of Mexico and Guatemala (Lowe, Ekholm and Clark 2013; Lowe, Lee and Martínez 1982), but the local organization of this polity was poorly

Figure 1. Map of Mesoamerica with sites mentioned in the text.

understood. In spite of this, Izapa's impressive earthen architecture and extensive corpus of carved monuments (Clark and Moreno 2007; Guernsey 2006 2016), led many scholars to speculate about the political power of the site's elite (e.g., Demarest 2004:67; Gómez Rueda 1996; Guernsey 2011; Lowe, Lee and Martínez 1982: 307; Love 2007: 291-292 2011). However, there has been no regional data from the region around Izapa with which to explore this notion. In fact, no knowledge of the political organization of the Izapa polity (or how the site's hinterland was structured) existed until the Izapa Regional Settlement Project (IRSP) was initiated in 2011.

In this paper, I review recent work by the IRSP that has documented the Izapa kingdom. I begin at a regional scale and define the Izapa kingdom as a system of centres with monumental earthen mounds, that lasted for approximately 600 years (700-100 BCE, all dates calibrated). Then, the eruption of the Tacaná volcano resulted in a decline in population as well as a reorganization of the political capital of Izapa between 100 BCE –100 CE. The second part of this paper describes the construction methods of the large earthen mounds at the Izapa capital city. New excavation data define a recently recognized E-Group at Izapa and the dating of this astrological architectural feature to the Escalón-Guillén phases (700-100 BCE). I then discuss how the capital was remade after the Tacaná eruption and a new pyramid group was built north of the Formative-period core. Excavation data from the site's large platform (Mound 30) reveal that it was doubled in size and raised in height during the Itstapa phase (100-300 CE).

## 2 Regional Kingdom of Izapa

The IRSP has collected lidar and pedestrian survey data that documents both regional population trends as well as the location and internal organization of over three dozen lower-order political centres dating to the Middle and Late Formative period (Rosenswig et al. 2013; Rosenswig and López 2018; Rosenswig, López and Antonelli 2015). The Izapa kingdom's territory is defined by

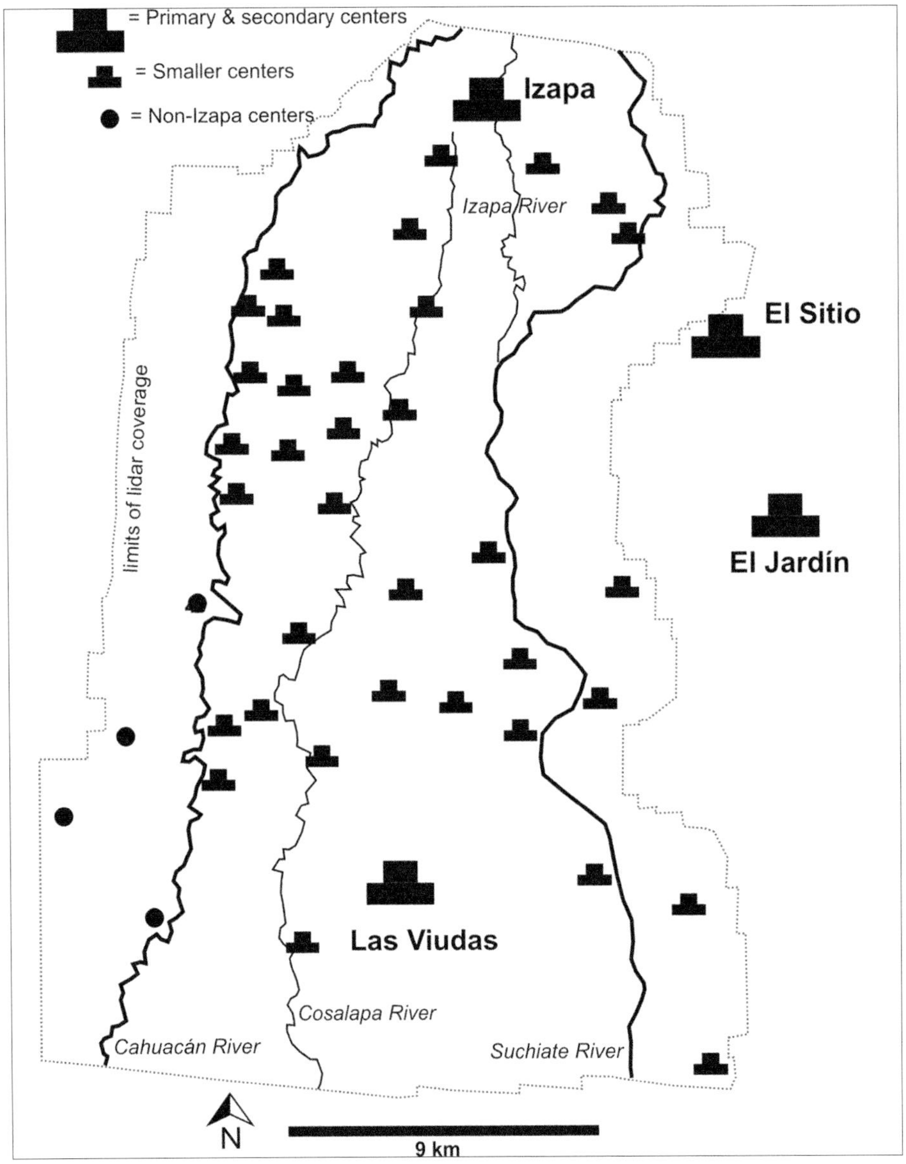


Figure 2. Map of the Izapa kingdom (based on Rosenswig and López 2018, fig. 3).

the Cahuacán River to the west and large centres on all other sides: the Izapa capital to the north and secondary centres of Las Viudas to the south, El Sitio and El Jardín to the east (Figure 2). Geographic barriers and population centres therefore provide a defensive perimeter around the kingdom's smaller centres. In the course of documenting so many lower-order centres, a number of consistent architectural features were identified. These are discussed in more detail by Rosenswig and López (2018) and together are referred to as being 'Izapa-pattern' sites. In brief, each centre has its largest conical temple mound to the north and, at primary, secondary and tertiary centres, this northern mound is on a square platform (Table 1).

South of the platform are two to four plaza groups formed by mounds that align south-southwest to north-northeast so that the overall orientation of each site is towards the volcano of Tacaná (as Izapa is), of the Tajumulco or points in between. The orientation of all centres is between 18-28 degrees east of magnetic north. The result is that at all 40 sites documented to date in the kingdom, from any of the plazas or mounds to the south, the northern pyramid and the 'stage' formed by the main temple and south end of the platform, are framed by the two volcanoes that rise dramatically from the Southern Sierra Madre mountain range. At Izapa, and all secondary centres, there is also an E-Group that defines the southern end of this architectural layout. I assume that the monumental architecture at the centre of each site reflects ceremonial and ritual beliefs and that it structured the daily activities of all residents. Furthermore, the similarity of

| Settlement Tier | Site Size (ha) | Qualitative Characteristics |
|---|---|---|
| 1 (Izapa) | 229 | Sculpture |
| | | E-Group/Ballcourt |
| | | 3 plazas |
| | | North platform |
| 2 | 14-43 | E-Group/Ballcourt |
| | | 3-4 plazas |
| | | North platform |
| 3 | 1-9 | 1-2 plazas |
| | | North platform |
| 4 | >1 | 1 plaza |

Table 1. Criteria for Izapa kingdom settlement hierarchy.

the urban layout within the kingdom of Izapa speaks of a set of ritual and social behaviour that both created and reflected a shared identity and some degree of cultural affiliation. As discussed in more detail by Rosenswig and López (2018), the similarity in architectural form and orientation within the Izapa kingdom contrasts with neighbouring polities of El Ujuxte and Takalik Abaj (see fig. 1) where different architectural standards reflect different ritual practices and thus different affiliations (Love 2007: 294-295). Within a common structure of kings and elite culture defining a shared hierarchical social structure across Pacific Mesoamerica's polities at this time (Love 2011), individual practices differentiated each kingdom. To be clear, the site of Izapa does not currently conform to the 'Izapa pattern' as the site was expanded repeatedly after its initial construction (Figure 3). Enough excavation has been conducted at the site to know it did conform to this pattern during the Middle Formative Escalón (700-500 BCE) and Frontera (500-300 BCE) phases, when Mound 30a defined the north end of the site and three plaza groups extended south in Group B and G (Clark and Lee 2013; Lowe, Lee and Clark 1982; Rosenswig et al. 2018). During the Guillén phase (300-100 BCE), Group A, D and other plaza groups were built to the west and additional in-filling of mounds during the Itstapa phase (100-300 CE) contribute to how the site is now formed (Lowe, Lee and Clark 1982; Rosenswig and Mendelsohn 2016). The city of Izapa was not static urban space. The kingdom emerged and grew and the capital evolved over many centuries. During the Guillén phase, this capital was distinguished from its lower-order centres in the kingdom both by building more plaza groups west of the original alignment and lining them with the low-relief, stone sculptures for which the site is famous.

### *2.1 Regional population*

The IRSP has defined regional patterns associated with the Formative-period Izapa kingdom. Three survey zones between the Cahuacán and Suchiate Rivers were systematically documented. Within the low hills and piedmont zones, lidar mapping documented all mounds measuring more than 50 cm in height and the periods of their occupation were determined on a phase-by-phase basis (Rosenswig et al. 2013; Rosenswig et al. 2015). In the coastal plain survey zone, hectares of occupation were also recorded on a phase-by-phase basis (Rosenswig 2008). Settlement data from these three survey zones provide the basis for the reconstructing of population changes in the region as well as the movement of majority of population from the coastal plain to the piedmont and low hills after 700 BCE (Figure 4). An additional campaign of lidar data collection, in 2015, brought the total coverage to almost 600 sq km and documented 40 lower-order monumental centres that were occupied during the Escalón, Frontera and Guillén phases (Rosenswig and López 2018). These regional data have enabled reconstruction of the internal spatial organization of the polity for the first time. It is now clear that, by the Escalón phase (700-500 BCE), Izapa was the capital of a regionally organized polity, consisting of dozens of monumental centres that each applied the same site planning principles (Blake, Rosenswig and Waber 2015; Rosenswig et al. 2015), and reflected a complex administrative hierarchy (Rosenswig and López 2018). Sites built according to the Izapa pattern possess mounds that form multiple plazas arranged in a roughly north-south alignment.

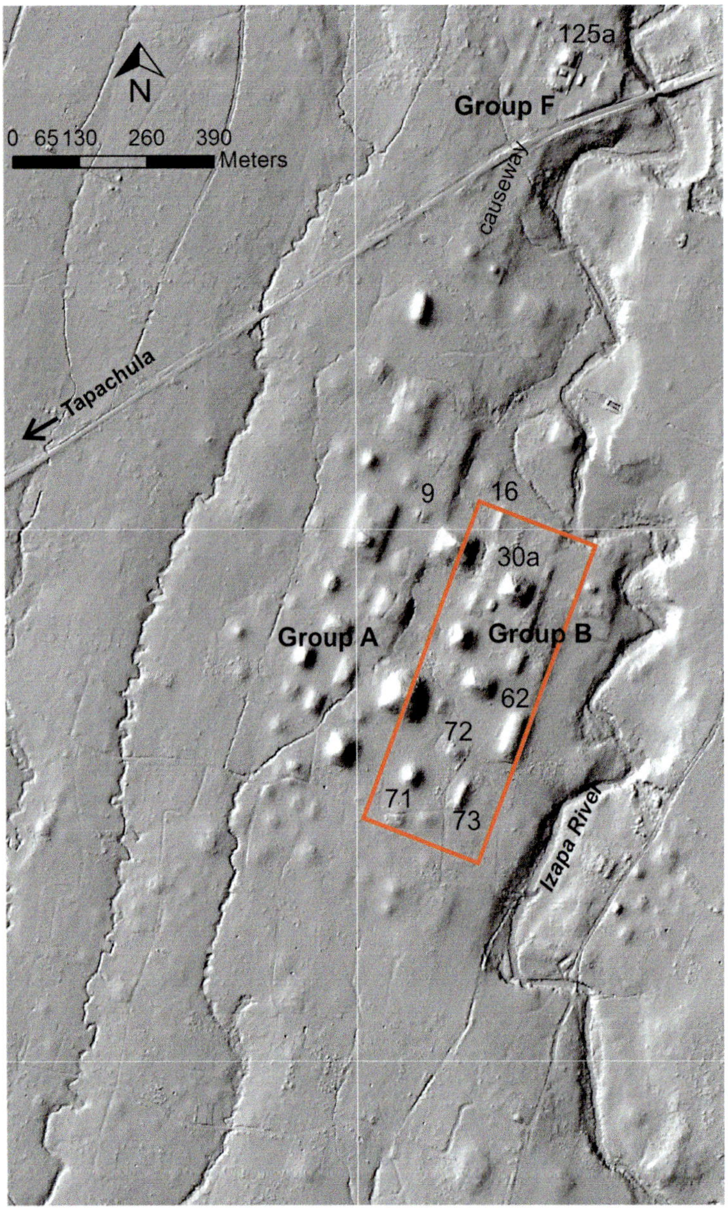


Figure 3. Hillshaded DEM of earthen architecture at Izapa with mounds mentioned in the text.

At each site's northern end sits a large pyramid atop a platform and, at the southern end, an E-Group. The pattern shares some characteristics with Clark and Hansen's (2001) Chiapas Middle Formative pattern (Lowe 1977).

Work by Neff et al. (2018) constitutes a complementary piece of the larger regional puzzle. Survey and excavations in the estuary south of Izapa by Neff and colleagues provide new evidence from this important resource procurement zone. Their work has documented human activity in the estuary by 4000 BCE and that after 1000 BCE, with the rise of the La Blanca polity (Love 2002; Rosenswig 2010), salt began to be harvested from this zone using the *sal cocida* technique of boiling down estuary water. Neff and colleagues documented firing features and distinct coarse-ware ceramics. The lack of a full assemblage of Middle Formative ceramic vessel forms, that include domestic wares, indicates that salt production was a specialized activity and that the estuary zone was an extension of the regional settlement focused inland. Salt consumption at this time must have increased due to the dietary requirements associated with increased reliance on maize (Rosenswig et al. 2015). Three to four times more sites were documented during the late Middle and Late Formative period Bermudez complex (450-100 BCE) than at any other time during the Formative era (Neff et al. 2018, fig. 5). Expanded salt production thus corresponds to the zenith of the Izapa kingdom. Occupation in the estuary then decreased significantly during the Terminal Formative Soledad phase (1-250 CE), when Izapa was experiencing volcano-induced political problems, as discussed below.

### *2.2 Eruption of Tacaná and reorganization of Izapa*

During the final century before the Common Era, after six centuries of relative stability, major changes occurred to the organisation of Izapa. As Lowe, Lee and Martínez (1982: 308) long-ago hypothesized, an eruption of the Tacaná volcano led to a disruption of Izapa's growth, which impacted the long-standing heart of the city in lower Izapa. Rather than a thriving political and economic centre, lower Izapa was transformed into a more sacred/religious space (Lowe 1993).

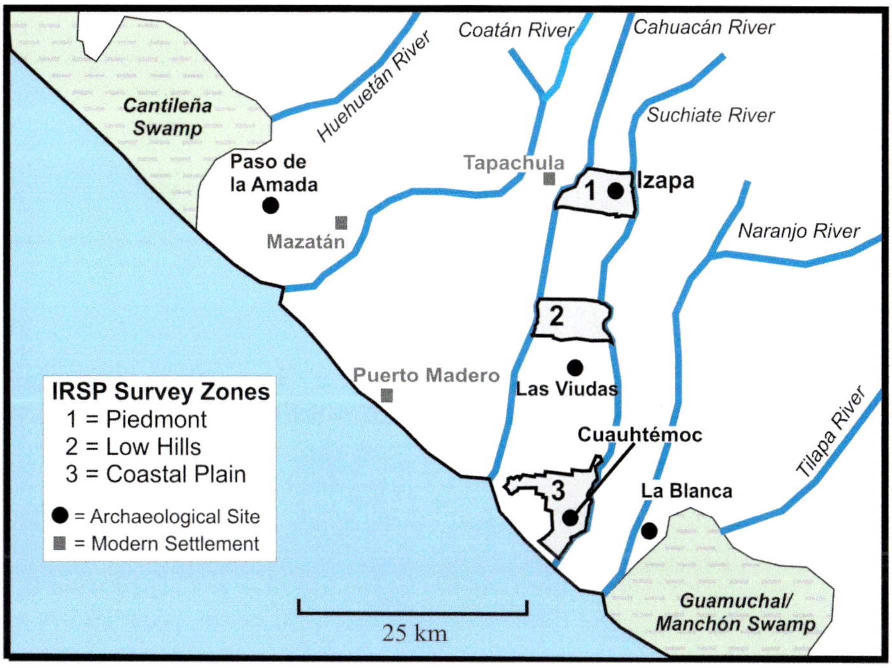


Figure 4. IRSP piedmont, low hills and coastal plain survey zones.

Rosenswig and Mendelsohn (2016, fig. 4a) have reported very limited occupation of lower Izapa during the Hato phase, which corresponds to the period in which construction began around Group F (see Figure 3). Macías et al. (2018, fig. 2a) present new radiometric dates associated with the eruption that have a modelled age range of between 30 BCE and 80 CE. This dating indicates that the end of the Guillén phase may correspond with this calamitous volcanic event. As Macías and colleagues describe, the pyroclastic flows descended to within 7 km of the Izapa centre and blanketed the site in ash, clogging all of the local rivers that soon would have overflowed and produced violent flooding.

Following this tumultuous event, the primary occupation at Izapa shifted north to Group F. The IRSP determined that mounds extending out from Group F covered as much as three times as large an area as had previously been reported by the NWAF (Rosenswig et al. 2013: 1504). This includes dozens of newly documented residential mounds and a large elite mound complex beyond what was mapped by the NWAF (Rosenswig and Mendelsohn 2016, fig. 3).

## 3 The Izapa Capital

The Izapa capital is located on the edge of a low piedmont below the Sierra Madre of Chiapas. The arrangement of Izapa's mounds were carefully documented and reported by the NWAF (Lowe, Lee and Martínez 1982, inset). IRSP remapping of Izapa's architecture using lidar technology revealed previously unrecognized features and determined the site's limits for the first time. Newly identified architectural features include formal plazas along the Izapa River and an E-Group at the site's south end (Rosenswig 2019; Rosenswig et al. 2013). Excavation data from Izapa indicate that Groups B and G were established in their current arrangement during the Escalón and Frontera phases (Lowe, Ekholm and Clark 2013; Lowe, Lee and Martínez 1982; Rosenswig et al. 2018). Mound construction was subsequently extended to the west and Groups A, D and H were built during the following Guillén phase.

The Duende phase (800-700 BCE) was when mounds were first built at Izapa and our results are consistent with NWAF excavation that document that Mound 30a was built up to 10 m above the platform during this time (Lowe, Lee and Martínez 1982). Mound 30a defines the northern end of Group B as the original centre of the site with the Izapa pattern. The Izapa capital reached its largest size during the Escalón phase with remains from this phase documented over an area of 229 ha. Using population estimates of 25 people/ha, this would represent a city with approximately 5725 inhabitants (see Rosenswig and López 2018). The following Frontera and Guillén phases have evidence of fewer hectares (189 and 156 ha respectively) over which remains were documented. However, it is unclear if this measure reflects fewer people living at the site or if similar numbers of people lived at a higher population density. The central core of the site (i.e., Groups B and G) was occupied during the Escalón, Frontera and Guillén phases as was a large area south of Group G. The change in number of hectares of occupation from the Escalón to the two later phases is due to occupation of less area north of the modern highway and west of the monumental core of lower Izapa. Based on IRSP results, occupation of Izapa fell to an area measuring 48 ha during the Hato phase (Rosenswig 2019). This is a third the size as during the previous Guillén phase and a fifth the size as during the Escalón phase. If the same estimate of 25 people/ha is used, the Hato phase population of Izapa would have been ~1200 people.

As discussed above, an eruption of the Tacaná volcano during the Hato phase brought the Mixcun lava flow to within 7 km of Izapa. Lahars associated with this lava may have also brought pyroclastic flows of hot, muddy water that could have been 6 m high as they coursed through the city (Macias et al. 2018). Such a natural disaster would have shaken the population's confidence in whoever ruled Izapa at the time, especially as the capital was oriented towards this volcano (Blake et al. 2015). Although the specific political details will never be known, monumental architecture ceased being constructed in the southern part of the capital. In addition, overall population levels fell during the Hato phase and the location that these people lived also changed.

Regional settlement patterns document contemporaneous disruptions were occurring throughout the Izapa kingdom during the first century BCE. During the Hato phase, there was a significant drop in overall population in both the piedmont and low hills zones of the IRSP survey (Rosenswig and Mendelsohn 2016, fig. 5). In the Mazatán zone, located 40 km to the northwest, there was a proliferation of sites during the Hato and Itstapa phases, after not a single site was documented from Guillén times. A lack of occupation in the Mazatán zone during the Guillén phase indicates that Izapa had exerted significant influence by inhibiting centres from developing (Clark and Lowe 2013: 84-85). Therefore, the disruption to the Izapa kingdom after 100 BCE was as dramatic regionally as at the capital city.

## 4 Izapa excavations

### *4.1 Middle and Late Formative E-Group*

An E-Group was recently recognized at Izapa based on the newly acquired IRSP lidar data (Rosenswig et al. 2013). First defined at Uaxactún (Blom 1924), E-Groups are identified as a pair of mounds – a conical one to the west and a long, linear one to the east – that together are located south of the earliest principal mound of a site. E-Groups have long been interpreted as serving astronomical functions (Aimers and Rice 2006; Clark and Hansen 2001) and the astrological alignments of Izapa's site layout is explored further elsewhere (Blake et al. 2015). In contrast to the way Mound 71 and 73 were originally depicted as two ovals (Lowe, Lee and Martínez 1982, inset), the lidar data document them in the shape and orientation expected for an E-Group (see fig. 3). Mounds 71 and 73 are located along the site's original Middle Formative axis directly south of Mound 30a. This is precisely where an E-Group would be expected to be.

The newly recognized E-Group was investigated with excavation during the 2012 season. Unfortunately, we were not granted permission to excavate at Mound 71. Permission was, however, granted to work at Mound 72 and 73 and we documented the late Middle Formative establishment of the E-Group and significant expansion during the Guillén phase. Two units excavated on the centre-line axis east and west of Izapa's Mound 73 (Figure 5). Suboperation 1a was excavated on

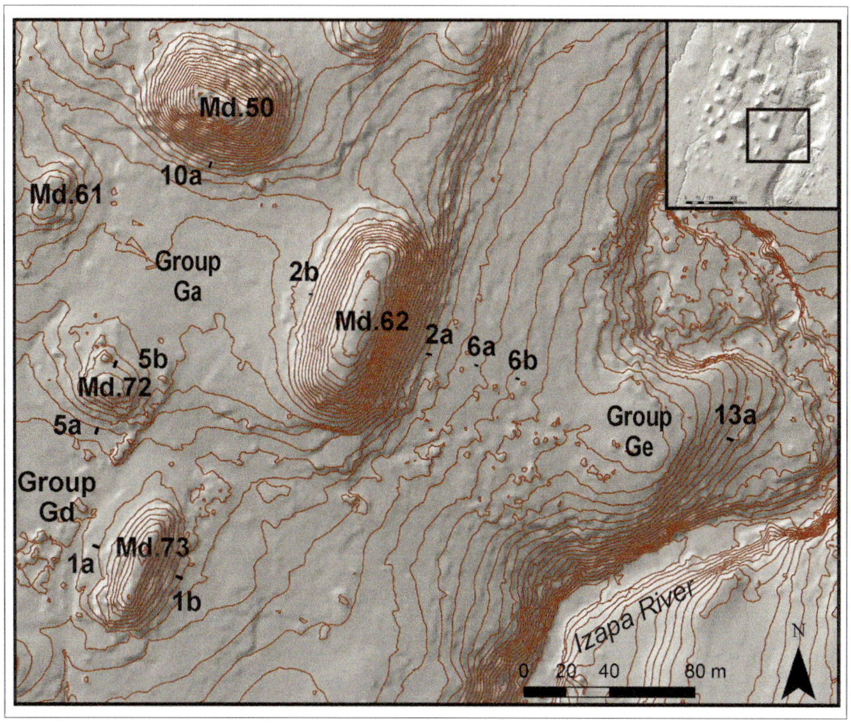


Figure 5. IRSP southern area of excavation at lower Izapa.

Figure 6. Suboperation 1b at Mound 73: Late Formative period earthen mound fill with Middle Formative period midden removed.

the west side of Mound 73 and in it, we documented the remains of a single episode of construction and ceramic sherds dating to the Late Formative Guillén phase and earlier. Suboperation 1b was excavated on the east side of Mound 73, and also documented a single episode of construction during the Late Formative period (Figure 6). Ash laid down by the 1902 eruption of Santa Marta in Guatemala is clearly visible at the top of the unit shown in fig. 6. Below this was a matrix that contained Guillén-phase (and earlier) ceramics that we interpret as Late Formative period construction fill. This layer was between 1 m and 1.4 m thick and consisted of dark brown clay of the same type documented in Suboperation 1a. Ceramic density from Suboperation 1b is more than twice as high as the construction fill documented in Suboperation 1a, and so, indicates that more cultural materials were mined to build the east side of Mound 73 than the west side that faced

Figure 7. Mound 72, photo faces east.

into Group Gd (Rosenswig et al. 2018, tab. 3). This is what we would expect if the plaza was kept relatively clean of debris and trash accumulated more on the west side of the mound and away from a formal plaza.

Below the Late Formative period construction fill in Suboperation 1b we documented an intact, Middle Formative midden. The midden was 20-40 cm thick and contained exclusively Escalón and Frontera phase ceramics, indicating that this locale was occupied for at least three centuries prior to the site's Guillén-phase apogee. The Middle Formative period midden layer was preserved by the overlying mound just above bedrock, as is shown in fig. 7, after we had removed it from the south and east walls of Suboperation 1b. A maize cupule recovered from this Middle Formative midden produced a date consistent with the established chronology of the Escalón phase with a 2-sigma result of 769-539 BCE (Rosenswig et al. 2018, fig. 7).

Excavations were also undertaken on the north and south sides of Mound 72 (Figure 7), which closes off the north side of Group G formed with Mounds 71 and 73. Excavations at this mound documented that it was first built during the Middle Formative period, and that later the south side was augmented during the Guillén phase. There is Itstapa phase debris in the uppermost levels in Suboperation 5a, on the south side of Mound 72. A meter below current ground surface, on the north side of Mound 72, a complete vessel cache was documented in Suboperation 5b within the stone covering the initial construction episode of the mound (Figure 8). A carbon sample collected 1 m south (i.e., towards the centre) and 30 cm lower than where the vessel was encountered, produced a 2-sigma result of 342-106 BCE. So, Mound 72, which lies just over 300 m directly south of Mound 30a (along the site's primary alignment), was built during the Escalón and Frontera phase and then augmented in Guillén-phase times.

Two of the mounds that form a plaza for Izapa's E-Group were built during the late Middle Formative and significantly augmented during the Guillén-phase apogee of the site. Duende-phase remains may be contained within the mounds' cores, but we have not yet penetrated them deeply enough to find out. In contrast, no evidence of Hato phase construction was present at either mound, indicating a cessation of construction at Izapa's E-Group by 100 BCE, which is consistent with IRSP

Figure 8. Suboperation 5b at Mound 72 showing Middle Formative core of the E-Group mound.

survey results discussed in the previous section.

### *4.2 Terminal Formative Mound 30 Platform Expansion*

Excavations were undertaken north of Group B (Figure 9). The northern half of the Mound 30 platform was built in three episodes during the Terminal Formative period. The construction episodes are visible as two lines of stone cobbles in the south wall profile of Suboperation 4a (Figure 10). Below the lower cobble surface, a dense midden was documented that contained Middle Formative ceramics. Between the two cobble surfaces is a dark brown matrix we interpret as construction fill that contains Terminal Formative Itstapa phase (and earlier) ceramic sherds, dating the fill to 100-300 CE. Above the top cobble surface is a light brown fill that also contains Terminal Formative sherds and extends up to the active root mat. Excavation results from Suboperation 3a, 4b as well as 9a and 9b (see Figure 9) provide the same stratigraphic pattern of three distinct construction episodes but not all are so clearly defined by intact stone cobbles as those from Suboperation 4a (Rosenswig et al. 2014).

The Terminal Formative enlargement of the size of the Mound 30 platform was considerable. The platform's extents almost doubled, and what was square during the Middle and Late Formative period, expanded to its current rectangular shape and size. The area of expansion, beyond the square platform on which Mound 30a sits, measures 150 x 150 m and is 2 m high. This translates to approximately 45,000 cu m of fill. The expenditure of so much effort is likely explained by the shift of the monumental focus of the site centre from lower Izapa to upper Izapa (i.e., Group F) that began during the Terminal Formative period. Lidar data have allowed us to recognize a causeway that began at Group F and extended 270 m south towards lower Izapa (Rosenswig and Mendelsohn 2016). If religiously and/or politically-motivated processions originated at Group F, and followed the causeway southward, then their point of entry (to what was by then Old Izapa) would have been Mounds 9 and 15 and the north edge of the Mound 30 platform (see Figure 3). Serving as an entry point to Old Izapa would explain why there was a modification to what would have been the 'back' of Mound 30a during the Middle and Late Formative period occupation of the site. A shift in Izapa's political centre to Group F, and processions periodically travelling to lower Izapa, would also explain why so many monuments were reset at Mound 9 after the older parts of Izapa (and Group B in particular) were no longer the focus of construction efforts at the site.

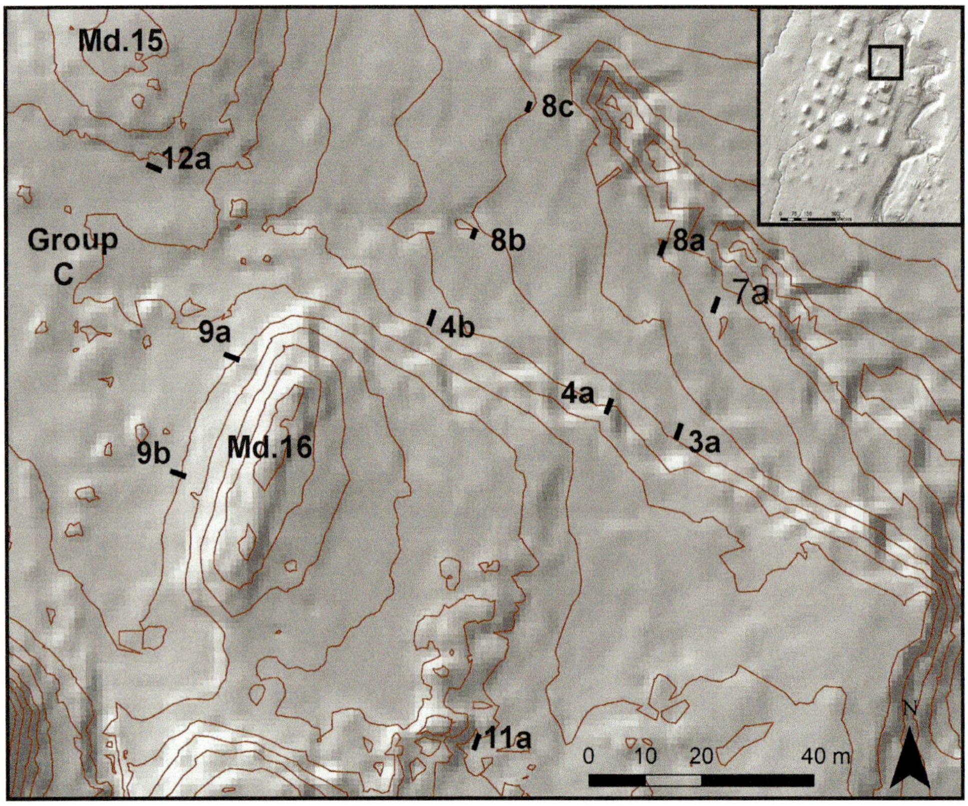


Figure 9. IRSP northern area of excavation at lower Izapa.

Figure 10. Suboperation 4a, south wall showing three construction phases of the Mound 30 platform.

Figure 11. Suboperation 4a, west wall showing profile of Terminal Formative period stone stairs in centre of northern edge of the Mound 30 platform.

Suboperation 4a was excavated at the middle of the north edge of the Mound 30 platform and, in addition to three episodes of Terminal Formative construction, also documents a central ramp and staircase. Figure 11 shows the south and west walls of Suboperation 4a where we encountered evidence of the stone cobble surfaces. Equivalent cobbles were not encountered in the east wall of Suboperation 4a. However, we did remove a number of large, flat-topped cobbles during the course of excavating this unit and thought they were part of the construction fill. It was only once they had been removed (and the unit's west wall became visible in profile), that we realized we must have removed the eastern-most edge of the ramp and staircase that terminated within the 1 m width of the excavation unit.

A single unit was excavated within the Mound 30 platform and further helps to date the platform's construction history. This excavation unit (Suboperation 11a) also provides stratigraphic association for Miscellaneous Monument 2 (aka El León), the oldest stone monument at the site inferred stylistically in terms of the three-dimensional nature of its carving and contextually for its association with Mound 30a (Navarette 2013: 19-21). Suboperation 11a was placed along the central axis of Mound 30a and 30 m north-northeast of this mound's north edge. This excavation unit was also 40 m west-northwest of El León. In Suboperation 11a, we documented a distinctive white clay layer, the top of which was 2 m below current ground surface (Figure 12). The height of the top of this clay layer is within 2 cm of that of the ground surface around the El León sculpture. The stratigraphy and associated radiocarbon assays documented in Suboperation 11a can thus be employed to date the depositional context of El León.

The white clay layer in Suboperation 11a was the result of multiple repavings of this ceremonial space during the Escalón phase. The south wall of Suboperation 11a shows three distinct episodes of white matrix deposition and a thin dark level between each of them (Figure 12). This indicates an initial construction and at least two subsequent resurfacing or maintenance episodes. The 'white' clay floors are actually grey (5Y 8/1) according to the Mussell standard. They were excavated as Levels 21 and 22 in Suboperation 11a. Within and below the clay surfaces, we recovered Escalón phase ceramic types (along with a few dating to the Duende phase). The first 30 cm below the floor in the south profile of Suboperation 11a (but as thick as 50 cm elsewhere in the unit) was a brown matrix. This layer contained a noticeably dense concentration of obsidian, ground stone and ceramics sherds during excavation. A 15-cm-thick level below contained the same matrix but also included a significant quantity of white clay. Below this, was a sandy orange brown matrix that also contained a high artefact density. The penultimate stratum was 15 cm thick and consisted of light

Figure 12. Suboperation 11a in centre of Mound 30 platform showing Late Formative plaster floor under Terminal Formative fill.

brown clay with numerous small stone inclusions and very few artefacts. The final 40 cm excavated in Suboperation 11a was sterile rock and sandy clay bedrock that is known locally as 'cascajo.'

Above the white clay floors documented in Suboperation 11a, we recovered significant quantities of Terminal Formative period ceramics mixed with earlier period types. We interpret this matrix as construction fill that dates to the Itstapa phase and extends to the north edge of the platform. Unlike the stratigraphy documented in Suboperation 4a and 4b, distinct construction episodes were not evident.

Ceramic analysis and radiocarbon assays along with stratigraphic evidence document that a late Middle Formative period platform was covered with construction fill during the Itstapa phase and that the Mound 30 platform reached its current size (see Rosenswig et al. 2018). The ceramic sherds presented in Figure 12 exemplify those recovered from the Terminal Formative period construction fill (documented in Levels 7 and 8) and those from below the floor are Middle Formative Escalón phase types (Levels 21 and 22). Two radiocarbon assays were recovered from these white clay surfaces. One was run from a maize cupule and kernel that produced a 2-sigma result of 355-120 BCE and the second was from a piece of charcoal embedded within the floor that dates to 729-410 BCE (Rosenswig et al. 2018, Table 2). We interpret the charcoal as dating the time the floor was first built (i.e., the Escalón phase) and the maize being from sometime during its later use (during the Frontera or possibly Guillén phase). Suboperation 11a provides hints of how the area north of Mound 30a was used by the residents of Izapa from Middle through Terminal Formative times. First as non-architectural occupation area north of Mound 30a during the Escalón and Duende phases, then as a paved surface through to the Guillén phase and finally as an expanded platform during the Itstapa phase. The Escalón (and possibly Duende) phase deposits below the clay floors provide a sealed context from the early establishment of the site as a monumental centre.

## 5 Discussion and conclusion

The Izapa kingdom's urban centres were arranged to maintain internal cohesion and external sovereignty. During its apogee between 700-100 BCE, the kingdom encompassed dozens of lower-order centres forming a four-tiered political hierarchy within 22 km of the capital. The Izapa capital's location on the edge of the piedmont, facilitated rapid canoe travel down-river to its subsidiary centres. Secondary centres were arranged around the perimeter of the kingdom territory protecting population at smaller sites from neighbouring kingdoms to the south and east. All Izapa cities were arranged and oriented according to the same principles, in contrast to neighbouring kingdoms of Takalik Abaj and El Ujuxte to the southeast and small independent centres on the northwest.

Elsewhere in Mesoamerica, similar political developments were underway at this time. In Oaxaca, the Monte Albán capital was defensively established on top of a 400-m-high plateau. The nearby political centre of San Martín Tilcajete expanded its size by drawing in more people and also laid out the monumental architecture that defined the capital differently than (and on an orientation that contrasted with) Monte Albán whose control was resisted from 500-300 BCE (Spencer and Redmond 2003). By the Late Formative period (300-100 BCE), Monte Albán consolidated the entire valley into an overarching state (Blanton et al. 1993). In contrast to the centralizing power of Monte Albán, none of the Pacific coast kingdoms dominated their neighbours during the Late Formative period. A number of roughly equivalently sized kingdoms created a détente where none were large enough to dominate.

The Izapa capital's elite experienced problems a few centuries later when the Tacaná volcano erupted during the Hato phase and the core of the monumental centre was no longer occupied. Other kingdoms in southern Mesoamerica, like El Ujuxte, Kaminaljuyu and Takalik Abaj, reached their political apogees at precisely this time (Love 2011; Inomata et al. 2014). The eruption of Tacaná around the beginning of the Common Era was a localized phenomenon with dramatic results only for the Izapa kingdom. It is quite possible that Izapa's troubles in the final century BCE benefited the nearby El Ujuxte and Takalik Abaj kingdoms.

The Izapa elite seems to have reinvented themselves during the Itstapa and Jaritas phases (100-400 CE), and built a new monumental core for their capital around Mound 125, north of the traditional centre. During these centuries of the early first millennium CE, the regional population also rebounded (Rosenswig and Mendelsohn 2016, Figure 5) and architectural alterations were undertaken at lower Izapa (Rosenswig et al. 2018). The two centuries from 200-400 CE was a time of collapse elsewhere in southern Mesoamerica (Love 2007; Inomata et al. 2014). Thus, Izapa seems to have been out of synch with political developments in the region and the kingdom's rulers charted their own distinct path.

## Bibliography

Aimers, J.J. and Rice, M. (2006) 'Astronomy, Ritual, and the Interpretation of Maya 'E-Group' Architectural Assemblages', *Ancient Mesoamerica*, 17(1): 79–96.

Blake, M., Rosenswig, R.M. and Waber, N. (2015) 'Izapa's Hinterland: The Use of Lidar Mapping to Examine the Layout and Spatial Orientation of Secondary Centers in the Soconusco Region, Chiapas, Mexico', Paper presented at the 80th Annual Meeting of the Society for American Archaeology, San Francisco.

Blanton, R.E., Feinman, G.M., Kowalewski, S.A. and Finsten, L. (1993) *Ancient Mesoamerica: A Comparison of Change in Three Regions*. 2nd edn. Cambridge: Cambridge University Press.

Blom, F. (1924) 'Report on the Preliminary Work at Uaxactun, Guatemala', *Carnegie Institution of Washington Yearbook*, 23: 217–219.

Chase, A.F., Chase, D.Z., Fisher, C.T., Leisz, S.L. and Weishampel, J.F. (2012) 'Geospatial revolution and remote sensing LiDAR in Mesoamerican archaeology', *Proceedings of the National Academy of Sciences* 109(32): 12916-12921.

Clark, J.E. and Hansen, R.D. (2001) 'The architecture of early kingship: comparative perspectives on the origins of the Maya Royal Court' in Inomata, T. and Houston, S.D. (eds) *Royal Courts of the Ancient Maya*, vol. 2. Boulder: Westview Press: 1-45.

Clark, J.E. and Lee, T.A. (2013) *Minor Excavations in Lower Izapa*. Papers of the New World Archaeological Foundation 75. Provo: Brigham Young University.

Clark, J.E. and Lowe, G.W. (2013) 'Izapa History' in Lowe, G.W., Ekholm, S. and Clark, J.E. (eds) *Middle and Late Preclassic Izapa: Ceramic Complexes and History*, Papers of the New World Archaeological Foundation 73. Provo: Brigham Young University: 69-87.

Clark, J.E. and Moreno, A. (2007) 'Redrawing the Izapa Monuments' in Lowe, L.S. and Pye, M.E. (eds) *Archaeology, Art, and Ethnogenesis in Mesoamerican Prehistory: Papers in Honor of Gareth W. Lowe*. Papers of the New World Archaeological Foundation 68. Provo: Brigham Young University: 277-319.

Clark, J.E. and Pye, M.E. (2000) 'The Pacific Coast and the Olmec Question' in Clark, J.E. and Pye, M.E. (eds) *Olmec Art and Archaeology in Mesoamerica*. Washington D.C.: National Gallery of Art: 217-251.

Cowgill, G. (2004) 'Origins and Development of Urbanism: Archaeological Perspectives', *Annual Review of Anthropology*, 33: 525-549.

Demarest, A.A. (2004) *Ancient Maya: The Rise and Fall of a Rainforest Civilization*. Cambridge: Cambridge University Press.

Evans, D.H., Fletcher, R.J., Pottier, C., Chevance, J.B., Soutif, D., Tan, B.S., Im, S., Ea, D., Tin, T., Kim, S., Cromarty, C., De Greef, S., Hanus, K., Bâty:, Kuszinger, R., Schimoda, I. and Boornazian, G. (2013) 'Uncovering archaeological landscapes at Angkor using lidar', *Proceedings of the National Academy of Sciences*, 110(31): 12595-12600.

Gómez Rueda, H. (1996) 'Izapa: Organización Espacial de un Centro del Formativo en la Costa Pacífica de Chiapas' in Laporte, J.P. and Escobedo, H.L. (eds) *IX Simposio de Investigaciones Arqueológicas en Guatemala 1995*. Guatemala City: Ministerio de Cultura y Deportes, Instituto de Antropología e Historia, and Asociación Tikal: 549-563.

Guernsey, J. (2006) *Ritual and Power in Stone: The Performance of Rulership in Mesoamerican Izapan-Style Art*. Austin: University of Texas Press.

Guernsey, J.(2011) 'Signifying Late Preclassic Rulership: Patterns of Continuity from the Southern Maya Zone' in Love, M. and Kaplan, J. (eds) *The Southern Maya in the Late Preclassic: The Rise and Fall of an Early Mesoamerican Civilization*. Boulder: University Press of Colorado: 115-138.

Guernsey, J. (2016) 'Water, Maize, Salt, and Canoes: An Iconography of Economics at Late Preclassic Izapa, Chiapas, Mexico', *Latin American Antiquity*, 27(3): 340-356.

Inomata, T., Ortiz, R., Arroyo, B. and Robinson, E.J. (2014) 'Chronological Revision of Preclassic Kaminaljuyú, Guatemala: Implications for Social Processes in the Southern Maya Area', *Latin American Antiquity*, 25(4): 377-408.

Kowalewski, S.A., Feinman, G.M., Finsten, L., Blanton, R.E. and Nicholas, L.M. (1989) *Monte Alban's Hinterland, Part II: Prehispanic Settlement Patterns in Tlacolula, Etla and Ocotlan, the Valley of Oaxaca, Mexico*. Memoirs, University of Michigan Museum of Anthropology 23. Ann Arbor: University of Michigan.

Love, M.W. (2002) *Early Complex Society in Pacific Guatemala: Settlements and Chronology of the Rio Naranjo, Guatemala*. Papers of the New World Archaeological Foundation 66. Provo: Brigham Young University.

Love, M.W. (2007) 'Recent Research in the Southern Highlands and Pacific Coast of Mesoamerica', *Journal of Archaeological Research*, 15(4): 275-328.

Love, M.W. (2011) 'City States and City-State Culture in the Southern Maya Region' in Love, M. and Kaplan J. (eds) *The Southern Maya in the Late Preclassic: The Rise and Fall of an Early Mesoamerican Civilization*. Boulder: Colorado University Press: 47-76.

Lowe, G.W. (1977) 'The Mixe-Zoque as Competing Neighbors of the Early Lowland Maya' in Adams, R.E.W. (ed.) *The Origins of Maya Civilization*. Albuquerque: University of New Mexico Press: 197-248.

Lowe, G.W., Ekholm, S.M. and Clark, J.E. (2013) *Middle and Late Preclassic Izapa: Ceramic Complexes and History*. Papers of the New World Archaeological Foundation 73. Provo: Brigham Young University.

Lowe, G.W., Lee Jr., T.A. and Martínez Espinoza, E. (1982) *Izapa: An Introduction to the Ruins and Monuments*. Papers of the New World Archaeological Foundation 31. Provo: Brigham Young University.

Macías, J.L., Arce, J.L., Capra, L., Saucedo, R. and Sánchez Núñez, J.M. (2018) 'Late Formative Flooding of Izapa after an Eruption of Tacaná Volcano', *Ancient Mesoamerica*, 29(2): 361-371.

Navarrete, C. (2013) *Excavaciones tempranas de Izapa*. Papers of the New World Archaeological Foundation, No. 75. Provo: Brigham Young University.

Neff, H., Burger, P.H., Culleton, B.J., Kennett, D.J. and Jones, J.G. (2018) 'Izapa's Industrial Hinterland: The Eastern Soconusco Mangrove Zone during Archaic and Formative Times', *Ancient Mesoamerica*, 29(2): 395-411.

Rosenswig, R.M. (2008) 'Prehispanic Settlement in the Cuauhtémoc Region of the Soconusco, Chiapas, Mexico', *Journal of Field Archaeology*, 33(4): 389-411.

Rosenswig, R.M. (2010) *The Beginnings of Mesoamerican Civilization: Inter-Regional Interaction and the Olmec*. New York: Cambridge University Press.

Rosenswig, R.M. (2019) 'The Izapa Kingdom's Capital: Formative-Period Settlement Patterns, Population and Dating Low-Relief Stelae', *Latin American Antiquity*, 30(1): 91-108.

Rosenswig, R.M., Culleton, B., Kennett, D., Lieske, R., Mendelsohn, R. and Núñez-Cortés, Y. (2018) 'The Early Occupation of Izapa: Recent Excavations, New Middle Formative Dating and Ceramic Analyses', *Ancient Mesoamerica*, 29(2): 373-393.

Rosenswig, R.M. and López-Torrijos, R. (2018) 'Lidar reveals the entire kingdom of Izapa, Mexico during the first millennium BCE', *Antiquity*, 92(365): 1292–1309.

Rosenswig, R.M., López-Torrijos, R. and Antonelli, C.E. (2015) 'Lidar Data and the Izapa Polity: New Results and Methodological Issues from Tropical Mesoamerica', *Archaeological and Anthropological Sciences*, 7(4): 487-504.

Rosenswig, R.M., López-Torrijos, R., Antonelli, C.E. and Mendelsohn, R.R. (2013) 'Lidar Mapping and Surface Survey of the Izapa State on the Tropical Piedmont of Chiapas, Mexico', *Journal of Archaeological Science*, 40(3): 1493-1507.

Rosenswig, R.M., and Mendelsohn, R.R. (2016) 'Izapa and the Soconusco Region, Mexico, in the First Millennium A.D.', *Latin American Antiquity*, 27(3): 357-377.

Rosenswig, R.M., Mendelsohn, R.R., Antonelli, C., Lieske, R. and Núñez Córtes, Y. (2014) *Proyecto de Reconocimiento Regional de Izapa 2012: Informe Técnico Parcial*. Mexico City: Consejo de Arqueología, Instituto Nacional de Antropología e Historia.

Rosenswig, R.M., VanDerwarker, A.M., Culleton, B.J. and Kennett, D.J. (2015) 'Is it Agriculture Yet? Intensified Maize-Use at 1000 cal BC in the Soconusco and Mesoamerica', *Journal of Anthropological Archaeology*, 40, pp- 89-108.

Sanders, W.T., Parsons, J. and Santley, R.S. (1979) *The Basin of Mexico: Ecological Processes in the Evolution of a Civilization*. New York: Academic Press.

Spencer, C.S. and Redmond, E.M. (2003) 'Militarism, Resistance, and Early State Development in Oaxaca, Mexico', *Social Evolution and History* 2(1): 25-70.

# Chapter 10

# Monumental earthen architecture in Teotihuacan, Mexico

Maria Torras Freixa
Independent researcher
maria.torrasfreixa@gmail.com

**Abstract**

Ancient urban settlements offer a framework to interpret human-environment interactions. The Mesoamerican city of Teotihuacan, c. 1-650 CE, is characterized by a carefully designed grid plan dominated by large-scale buildings. Erected contemporaneously (200-250 CE), both the Sun Pyramid and Building 4 of the Moon Pyramid are examples of earthen architecture covered with stone and lime plaster facades. The monumentalization process of the city's core had a high impact on the local environment. Thus the study of earthen architecture can afford insights into political evolution, urban planning and Teotihuacan's interaction with its hinterland.

**Keywords:** Stamped earth, Construction cells, Urban planning, Rulership, Human-environment interactions

**Résumé**

Les sites urbains anciens offrent un cadre adéquat pour comprendre les interactions humain-environnement. La cité mésoaméricaine de Teotihuacan, c. 1-650 CE, se caractérise par son plan urbain quadrillé dominé par des monuments de grandes dimensions. Édifiés au même moment, la Pyramide du Soleil et le Bâtiment 4 de la Pyramide de la Lune sont des exemples d'architecture de terre avec des façades revêtues de pierre et de stuc. Le processus de monumentalisation du centre cérémoniel eut un fort impact sur l'environnement local. Ainsi, l'architecture de terre nous permet d'étudier l'évolution politique de Teotihuacan, son tracé urbain et les interactions avec son hinterland.

**Mots-clés:** terre battue, caissons de remblai, planification urbaine, gouvernement, interactions humain-environnement

## 1 Introduction

The tradition of earthen architecture in the Central Mexican Highlands has begun to be investigated only recently, for example the pre-Hispanic settlements of Cholula (Pérez 2016) and Tlalancaleca (López and Kabata 2018; Murakami et al. 2017, fig. 9), as well as the ancient city of Teotihuacan now included in the World Heritage List (UNESCO 2012). This archaeological site, however, is best known for its stone construction technology. Although there are some studies that address building techniques and construction materials used in Teotihuacan (Marquina 1951; Margáin 1966; Cabrera 1991; Morelos 1993; Barba and Córdova 2010; Murakami 2010), few data and observations have been published so far of the earthen buildings, while analyses of earth as building material are scarce and focus mainly on paleoenvironmental aspects (Rivera et al. 2007; Sánchez et al. 2013).

The pre-Columbian urban centre of Teotihuacan, dating from the period c. 1-650 CE, is situated in the northeastern sector of the Basin of Mexico, in the Teotihuacan Valley, about 45 km from modern-day Mexico City. This settlement was unique in contemporary Mesoamerica for its size (125,000 inhabitants in an area of 20 km2) and its perfectly designed grid layout, dominated by the Sun and Moon Pyramids (Millon 1973; Cowgill 2015) (Figure 1). During the Early Tlamimilolpa phase (200-250 CE), these two civic-ceremonial structures were built in their monumental form (Sugiyama and Cabrera 2007; Sugiyama 2013 2017; Sugiyama et al. 2013). Both the Sun Pyramid and Building 4 of the Moon Pyramid are examples of earthen buildings covered with stone and faced with lime plaster. Scholars have identified the soil employed in the fill of these buildings as previously used in agriculture (Manzanilla 2005; Rivera et al. 2007; Barba and Córdova 2010: 120; Sánchez et al. 2013).

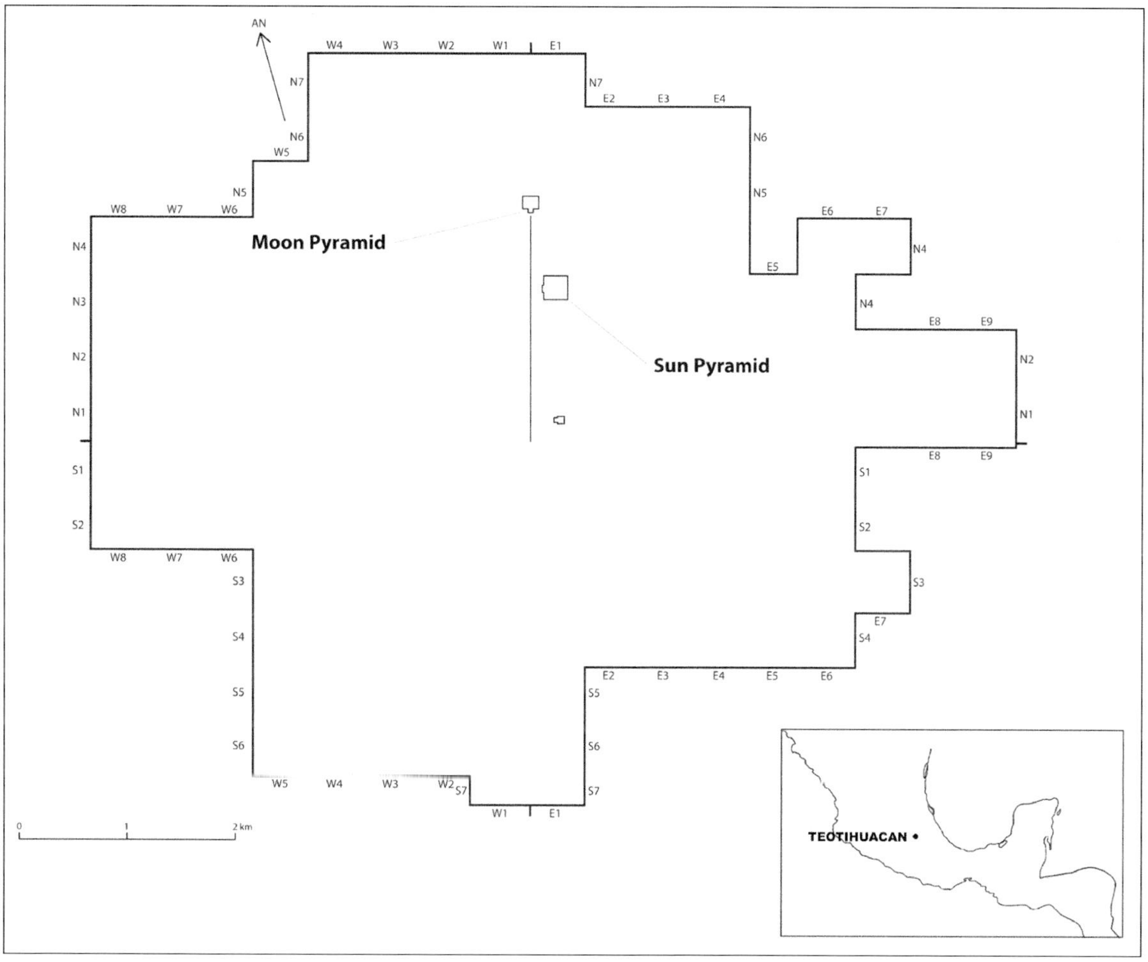


Figure 1. Plan of the city of Teotihuacan showing the location of the monumental buildings discussed in the text (modified after Millon 1973).

This short paper proposes to review published data on technological, paleobotanical, and paleoenvironmental factors of these two contemporaneous large-scale earthen buildings: the Sun Pyramid and Building 4 of the Moon Pyramid. The primary aim of this contribution is to show the importance of earthen architecture in the monumentalization process of the city's core, and the secondary objective is to explore the implications of this choice of building material, as it provides insights into sustainability and human-environment interactions in Teotihuacan.

## 2 Large-scale earthen buildings

The civic-ceremonial centre of Teotihuacan achieved its monumentality over a very short period of time (200-250 CE), involving a huge input of labor. Three monumental pyramids dominated the city's core: the Sun Pyramid, the Moon Pyramid and the Feathered Serpent Pyramid. Each architectural unit had its own building sequence, and each was built with different materials and techniques. The Feathered Serpent Pyramid is a square stepped pyramidal structure with tier in *talud-tablero* (a combination of sloping wall and vertical panel), and it was built with construction cells filled with stones and mud combined with wooden posts (Cabrera 1991). As volcanic rocks predominated for cells and fills (Murakami 2010: 167), this temple platform will not be reviewed in this paper. Nevertheless, the Sun Pyramid and Building 4 of the Moon Pyramid were erected using earth-based materials; thus they represent interesting examples of this architectural tradition. Also, this type of large-scale building provides information that can deepen our understanding of the city's urban planning and its impact on the environment.

Figure 2. Main façade of the Sun Pyramid (photograph by Maria Torras Freixa).

The first case is the Sun Pyramid, the largest and most impressive construction at the site. This iconic monument is a square stepped pyramidal structure with *taludes* (the tiers have sloping walls). Its first stage measured 216 m at the base and 64 m in height, and the volume of the entire structure was approximately 1,400,000 m3 (Millon and Drewitt 1961; Barba and Córdova 2010: 67). For a long time considered to belong to the Tzacualli phase (1-150 CE), the earliest building stage in Teotihuacan, until the $^{14}C$ results of the Sun Pyramid Project (2008-2011) revealed a new chronology for the pyramid at around 200 CE (modelled by Bayesian statistics 1 sigma 170-310 CE) (Sugiyama et al. 2013).

In many architectural and archaeological publications, the Sun Pyramid is described as an example of adobe construction (Millon and Drewitt 1961; Millon, Drewitt and Bennyhoff 1965) or even rammed earth (*tapial*), while in fact it was built with a stamped earth technique (Daneels 2015). From a central point marked by stones, this pyramidal structure was erected in a single episode by mounding soil directly above the bedrock (Sugiyama N. et al. 2013, Sugiyama S. et al. 2014) (Figure 2). The fill of the main body is primarily made up 'of a mixture of soil, ground tepetate [volcanic tuff], and silty sand layers with high concentration of charcoal and cultural remains' with hardly any stone or adobe bricks (Sugiyama et al. 2013: 429). In addition, 'the presence of artefacts and carbon fragments in the fill suggests that it was brought in from surface debris' (Sugiyama et al. 2013: 410).

The second case is the Moon Pyramid, starting point for the Avenue of the Dead, the main axis of Teotihuacan's grid plan. Excavations made by the Moon Pyramid Project (1998-2004) into the core of this monumental structure revealed seven superimposed building stages (Sugiyama 2004; Sugiyama and Cabrera 2006 2007). The first platforms, Buildings 1 to 3, are very modest in size and their fills are a mix of stones and earth. Erected around 250 CE, Building 4 shows a marked change in its dimensions, construction techniques and building materials. It is a square stepped pyramidal structure with tiers in *talud*, growing to nine times the size of the former three platforms and measuring around 89 m at its base (Sugiyama 2004; Sugiyama and Cabrera 2007). The fill of Building 4 is mainly earth, probably in a cell system (Murakami 2010: 167). More specifically, archaeologists have identified earth, adobe bricks and *tepetate* (volcanic tuff) -crushed or in blocks- among the construction materials used to enlarge the previous mound built of earth and rocks (Building 3) (Sugiyama and Cabrera 2000; Murakami 2010: 167; Sugiyama et al. 2013). Thus, a control method for the internal pressure of the fill using construction cells (Daneels 2015; Daneels et al. 2018) can be observed in Building 4 (Figure 3).

An interesting topic in monumental architecture is the duration of construction. Murakami (2010 2015) has developed this research in Teotihuacan through labor-input analysis and replicative experiments. In brief, he argued that each architectural unit, including the Sun Pyramid, was

Figure 3. View of the Moon Pyramid from the south-west of the Moon Plaza (photograph by Maria Torras Freixa).

erected in less than 10 years. In order to establish how long it took to build these architectural units, he estimated total labor costs (person-days) and per capita labor costs (days per person). He assumes that one person from every five-member household participated in building the central core, devoting 30, 60 or 100 workdays per year to construction. Also, following Abrams (1994), Murakami posits a five-hour workday for procurement and transportation and an eight-hour day for manufacture and assembly. Thus, energetic calculations (Murakami 2010 2015) support the chronological view that the Sun Pyramid and Building 4 of the Moon Pyramid were built contemporaneously or one slightly after the other in a very brief time (200-250 CE).

Using both micromorphology and botanical analysis (of macro-remains, pollen, phytolith and diatoms), scholars have studied some samples from the sediments making up the pyramids' construction fills. Results show that before being used as building material, the soil had been cultivated and irrigated (Sugiyama 2004; Gama et al. 2005; Manzanilla 2005; Rivera et al. 2007; McClung and Barba 2011; Sánchez et al. 2013; McClung 2015; Sánchez 2015). Specifically, the soil on which Teotihuacan is set has been identified as Black San Pablo Paleosol (BSPP) and it has been recovered at the Moon Pyramid fills (Rivera et al. 2007; Sánchez et al. 2013). BSPP is classified as a calcic Vertisol and displays a very dark organic horizon, characterized by high clay and organic matter content (Solleiro et al. 2011; Sánchez et al. 2013; Sánchez 2015).

## 3 Discussion

As described above, the Sun Pyramid and Building 4 of the Moon Pyramid are the largest earthen pyramidal structures in the ancient city of Teotihuacan. These two monumental buildings share certain features: for example, they were both erected on the bedrock (tepetate layer), their construction fills were made of soil removed from the surrounding area, their surface covered with stones faced with lime plaster, and they date from the same chronological range (200-250 CE). However, their ancient builders used different techniques in the fill of the pyramids: stamped earth (Sun) or construction cells mainly made of adobe and tepetate blocks (Moon). Writing of construction technology, Daneels (2015) suggests that stamped earth is more common on the Atlantic watershed and adobe bricks on the Pacific. Nevertheless, she adds that both techniques were used in the Mesoamerican tradition due to the geography and cultural interactions of the region (Daneels 2015).

An explanation of this technological dissimilarity may be found in the construction sequence of the city and its urban development. In the subsequent chronological phases, monumental architecture such as Buildings 5 to 7 of the Moon Pyramid and the Feathered Serpent Pyramid were erected with a cell system, but using stones and rocks also, this becoming a characteristic trait of Teotihuacan architecture (Murakami 2010: 167). Thus the chronological order of the erection of the two monuments -the Sun Pyramid slightly earlier than Building 4 of the Moon Pyramid- may explain their construction differences, although this can only be a preliminary hypothesis (Torras 2018). However, it is not a uniform construction sequence because some small temple platforms, such as Group 5', were built with adobe construction cells belonging to the Miccaotli (150-200 CE) and Early Tlamimilolpa phase (200-250 CE) (Paz 1995; Daneels et al. 1998). Better comprehension of the evolution of earthen architecture throughout the history of Teotihuacan may well yield information on the use and choice of different building techniques.

From the dawn of the urban settlement to its eclipse, the choice of construction materials is also an important factor to consider. In this context, the decision to use agricultural earth for large-scale architectonic units may have been a political choice, helping consolidate the layout and formal composition of Teotihuacan's civic-ceremonial core. As Barba writes, the pyramid fill material was obtained locally by razing adjacent areas of topsoil on a surface of 5 km2 and 40 cm deep to the bedrock (Barba and Córdova 2010; McClung and Barba 2011). This area in particular corresponds to the civic-ceremonial core. In this sense, the intentional destruction of this area transformed the built environment and gave more visibility to the existent sacred places through a process of monumentalization. Firstly, the choice of using agricultural soil as construction material may have had a symbolic meaning, characterizing the Sun Pyramid as a sacred mountain, *tonacatepetl* or mountain of abundance [in food] (Manzanilla 2005; Barba and Córdova 2010: 23, 149). Secondly, the pyramids were located in the same places where artificial tunnels and cavities had been dug during the Tzacualli phase (1-150 CE), thus emphasising the importance of these inner spaces for the society of Teotihuacan. Furthermore, respect for the previous locations of public spaces reflects the consolidation of a broader city plan (Torras 2018 2019). Thus, large-scale earthen buildings contributed to developing a common shared cultural identity and a worldview that shifted from the underworld to the terrestrial level (Torras 2018 2019).

One of the values of earth as a building material is its sustainability. Nevertheless, in the case of Teotihuacan, erecting these monumental buildings, along with the abandonment of irrigation systems due to the urbanization process of the whole city, had an environmental impact, directly affecting the city's agricultural sustainability (Torras 2019). Besides the consolidation of city's core, the sudden explosion of monumentality shows the intentional large-scale destruction of agricultural resources, razing BSPP and cultivated soil, for urban purposes (Gama et al. 2005; Rivera et al. 2007; Barba and Córdova 2010; McClung and Barba 2011; McClung 2012; Sánchez et al. 2013; Sánchez 2015). Thus the selection of earth as a raw material had a huge effect on the landscape, distinctly modifying human-environment interaction in the Teotihuacan Valley. In this context, monumental earthen buildings reveal valuable information regarding political processes. Although the carrying capacity of maize-based agricultural production in the Teotihuacan Valley was already exceeded at the end of the Tzacualli phase c. 150 CE (McClung 1990 2012), the materialization of ideology through large-scale buildings was more important for the government of the city. Also, the expansion of the built environment to the detriment of agricultural production seems to support the thesis of an empowerment of rulership and a successful control over new food supply, since the city continued to grow until 550 CE (McClung and Barba 2011, McClung 2012; Torras 2018).

How Teotihuacan society achieved this is still unclear; however some scholars suggest that new strategies were developed to obtain subsistence products from a symbiotic region in Central Mexico that included the Basin of Mexico, the Toluca Valley and the Puebla-Tlaxcala region (Sanders,et al. 1979; McClung 1987 2012). In addition, García, Gamboa and Vélez (2015) have documented an increase in rural settlements located near agricultural land or raw materials (such as basalt, andesite or *tepetate*) in the Basin of Mexico during the Miccaotli-Tlamimilolpa phase (150-350 CE). Thus, the rise to monumentality of the city matches the growth of the rural area. However, how the rulers of Teotihuacan interacted with the surrounding region requires further investigation.

As Blanton et al. (1996) have argued, political strategies can be reflected by monumental buildings. The economic and political decision-making necessary to build these large-scale earthen pyramids in so brief a time points to a growth in top-down control of the city and the existence of a powerful rulership. The government of Teotihuacan was able to shape the urban environment, manage a large labour force and expand the city's rural hinterland. In short, while the carrying capacity of the valley was exceeded, the city's political authorities were strong enough to offset this disadvantage in their bid to increase their authority and the material embodiment of their power. Based on this evidence, it seems plausible that the city's government achieved more power at local (the Teotihuacan Valley) and regional (the Basin of Mexico) levels, along with some kind of control in the wider area of ancient Central Mexico.

## 4 Conclusions

This study shows that in-depth (i.e., not solely descriptive) characterization of the construction materials and technological procedures of earthen architecture is essential. Each monumental earthen pyramid was built with different techniques, but this is not well understood. Analyzing these dissimilarities may well contribute to a better understanding of Teotihuacan's chronology and the development of its architecture. Thus Teotihuacan's large-scale earthen buildings should be studied further, using mechanical, mineralogical and chemical analysis.

The analysis of monumental earthen architecture offers a research framework for understanding human-environment interactions, as such constructions had a high impact. In the case of Teotihuacan, the use of agricultural soils as a source of construction material, must have changed the relations with the population of the city and the surrounding areas.

In addition, the choice of earth as a building material suggests an increase in the power of a central authority capable of transforming the city: the pyramids shaped the urban core and consolidated the city's grid layout, implementing a common worldview, managing labour investment, and ensuring a supply network robust enough to risk extending the urban settlement at the expense of agricultural land.

## Bibliography

Abrams, E.M. (1994) *How the Maya built their world: energetics and ancient architecture.* Austin: University of Texas Press.

Barba, L. and Córdova, J.L. (2010) *Materiales y energía en la arquitectura de Teotihuacan.* México: IIA, Universidad Nacional Autónoma de México.

Blanton, R.E., Feinman, G., Kowalewski, S. and Peregrine, P. (1996) 'A Dual-Processual Theory for the Evolution of Mesoamerican Civilization', *Current Anthropology*, 37(1): 1–14.

Cabrera, R. (1991) 'Los sistemas de relleno en algunas construcciones teotihuacanas' in Cabrera, R., Rodríguez, I. and Morelos, N. (eds) *Teotihuacan 1980-1982. Nuevas interpretaciones.* México: Instituto Nacional de Antropología e Historia: 113–143.

Cowgill, G. (2015) *Ancient Teotihuacan. Early Urbanism in Central Mexico.* New York: Cambridge University Press.

Daneels, A. (2015) 'Los sistemas constructivos de tierra en el méxico prehispánico' in *Seminario Iberoamericano de Arquitectura y Construcción con Tierra. 15o SIACOT -Ecuador.* Cuenca: Universidad de Cuenca, PROTERRA: 1–13.

Daneels, A., Domenici, D., Goldsmith, K., Grazioso, L., Layet, V., Moragas, N., Nobile, J.C., Paz, C. and Rodríguez, V. (1998) 'The architectural development of a three-temple complex. Excavations in Group 5', Teotihuacan, Mexico' in Bermond, G., Francovich, R., Mori, F., Pensabene, P., Salvatori, S., Tosi, M. and Peretto, C. (eds) *Proceedings of the XIII International Congress of Prehistoric and Protohistoric Sciences, Forlì, 8-14 September 1996*. Forlí: A.B.A.C.O Edizioni: 487-492.

Daneels, A., Love, M. and Beltrán, E.A. (2018) 'Control de presión interna de rellenos en basamentos de la Mesoamérica prehispánica' in Neves, C., Salcedo, Z. and Borges, O. (eds) *Memorias del Seminario Iberoamericano de Arquitectura y Construcción con Tierra. 18o SIACOT -Guatemala*. La Antigua Guatemala: USAC-CII, PROTERRA: 36-45.

Gama-Castro, J.E., Solleiro-Rebolledo, E., McClung, E., Villalpando, J.L., Sedov, S., Jasso-Castañeda, C., Palacios-Mayorga, S. and Hernández, D. (2005) 'Contribuciones de la Ciencia del Suelo a la investigación arqueológica: el caso de Teotihuacan', *Terra Latinoamericana*, 23(1): 1–11.

García Chávez, R., Gamboa, L. and Vélez, N. (2015) 'Los sitios rurales y la estrategia expansionista del estado teotihuacano para la captacion de recursos en la Cuenca de México', *Ancient Mesoamerica*, 26(2): 423–442.

López Juárez, J.M. and Kabata, S. (2018) 'La tecnología arquitectónica y su materialización en Tlalancaleca, Puebla', paper presented at the ICA 56th International Congress of Americanists, Salamanca, Spain, 15-20 July 2018.

Manzanilla, L. (2005) 'La construcción del inframundo en el Centro de México' in Pacheco, V. (ed.) *IV Coloquio Pedro Bosch Gimpera. El Occidente y Centro de México. I*. México: IIA, Universidad Nacional Autónoma de México: 273–302.

Margáin, C.R. (1966) 'Sobre sistemas y materiales de construcción en Teotihuacan' in *Teotihuacan Onceava mesa redonda*. México: Sociedad Mexicana de Antropología: 157–212.

Marquina, I. (1951) *Arquitectura prehispánica*. México: Instituto Nacional de Antropología e Historia, Secretaría de Educación Pública.

McClung, E. (1987) 'Patrones de subsistencia urbana en Teotihuacan' in McClung, E. and Rattray, E. (eds) *Teotihuacan. Nuevos datos, nuevas síntesis, nuevos problemas*. México: Universidad Nacional Autónoma de México: 57–74.

McClung, E. (1990) 'Modelos para la reconstrucción de actividades de subsistencia en la Cuenca de México durante el Clásico' in Cardós de Méndez, A. (ed.) *La época clásica: nuevos hallazgos, nuevas ideas*. México: Instituto Nacional de Antropología e Historia, MNA: 107–114.

McClung, E. (2012) 'Silent Hazards, Invisible Risks: Prehispanic Erosion in the Teotihuacan Valley, Central Mexico' in Cooper, J. and Sheets, P. (eds) *Surviving Sudden Environmental Change Understanding Hazards, Mitigating Impacts, Avoiding Disasters*. Boulder: University of Colorado Press: 143–165.

McClung, E. (2015) 'Holocene Paleoenvironment and Prehispanic Landscape Evolution in the Basin of Mexico', *Ancient Mesoamerica*, 26(2): 375–389.

McClung, E. and Barba, L. (2011) 'Ciencias aplicadas al estudio de estructuras monumentales en Teotihuacan, México', *Archaeological Practice: A Journal of the Society for American Archaeology*, (January): 19–28.

Millon, R. (1973) *The Teotihuacan map, Vol. I and II*. Austin: University of Texas Press.

Millon, R. and Drewitt, B. (1961) 'Earlier Structures within the Pyramid of the Sun at Teotihuacan', *American Antiquity*, 26(3): 371–380.

Millon, R., Drewitt, B. and Bennyhoff, J.A. (1965) 'The Pyramid of the Sun at Teotihuacán: 1959 Investigations', *Transactions of the American Philosophical Society*, 55(6): 1–93.

Morelos, N. (1993) *Proceso de producción de espacios y estructuras en Teotihuacán*. México: Instituto Nacional de Antropología e Historia.

Murakami, T. (2010) Power relations and urban landscape formation: a study of construcion labor and resources at Teotihuacan. PhD Thesis. Tempe: Arizona State University.

Murakami, T. (2015) 'Replicative construction experiments at Teotihuacan, Mexico: Assessing the duration and timing of monumental construction', *Journal of Field Archaeology*, 40(3): 263–282.

Murakami, T., Kabata, S., López, J.M. and Chávez, J.J. (2017) 'Development of an early city in Central Mexico: the Tlalancaleca Archaeological Project', *Antiquity*, 91(356): 455-473.

Paz Bautista, C. (1995) 'El Grupo 5', conjunto de tres templos Miccaotli-Tlamimilolpa temprano en Teotihuacan', *Revista Mexicana de Estudios Antropológicos*, 41: 101-115.

Pérez Castellanos, N. (2016) Los adobes arqueológicos de la Gran Pirámide de Cholula: caracterización detallada de su composición relacionada con sus propiedades como material de construcción. PhD dissertation. Mexico: Universidad Nacional Autónoma de México.

Rivera-Uria, M.Y., Sedov, S., Solleiro-Rebolledo, E., Pérez-Pérez, J., McClung, E., Gonzalez, A. and Gama-Castro, J. (2007) 'Degradación ambiental en el valle Teotihuacan: evidencias geológicas y paleopedológicas', *Boletín de la Sociedad Geológica Mexicana*, 59(2): 203–217.

Sánchez-Pérez, S., Solleiro-Rebolledo, E., Sedov, S., McClung, E., Golyeva, A., Prado, B. and Ibarra-Morales, E. (2013) 'The Black San Pablo Paleosol of the Teotihuacan Valley, Mexico: Pedogenesis, fertility, and use in ancient agricultural and urban systems', *Geoarchaeology*, 28(3): 249–267.

Sánchez-Pérez, S. (2015) Los paleosuelos 'negros' como indicadores de cambios ambientales naturales e inducidos por el hombre el periodo de ocupación teotihuacana. PhD Thesis. Mexico: Universidad Nacional Autónoma de México.

Sanders, W.T., Parsons, J.R. and Santley, R.S. (1979) *The Basin of Mexico. Ecological processes in the evolution of a civilization.* New York: Academic Press.

Solleiro-Rebolledo, E., Sycheva, S., Sedov, S., McClung, E., Rivera-Uria, Y., Salcido-Berkovich, C. and Kuznetsova, A. (2011) 'Fluvial processes and paleopedogenesis in the Teotihuacan Valley, México: Responses to late Quaternary environmental changes', *Quaternary International*, 233: 40–52.

Sugiyama, N., Sugiyama, S. and Sarabia, A. (2013) 'Inside the sun pyramid at Teotihuacan, Mexico: 2008-2011 excavations and preliminary results', *Latin American Antiquity*, 24(4): 403–432.

Sugiyama, S. (2004) *Viaje al Centro de la Pirámide de la Luna. Recientes descubrimientos en Teotihuacan.* México: CONACULTA-Instituto Nacional de Antropología e Historia, Arizona State University.

Sugiyama, S. (2013) 'Creation and transformation of monuments in the ancient city of Teotihuacan', in Sugiyama, S., Kabata, S., Taniguchi, T. and Niwa, E. (eds) *Constructing, deconstructing and reconstructing social identity: 2000 years of monumentality in Teotihuacan and Cholula, Mexico.* Aichi: Aichi Prefectural University: 1–10.

Sugiyama, S. (2017) 'Teotihuacan: planned city with cosmic pyramids' in Robb, M.H. (ed.) *Teotihuacan: City of water, City of fire.* San Francisco: Fine Arts Museum of San Francisco, University of California Press: 219-242.

Sugiyama, S. and Cabrera, R. (2000) 'El Proyecto Pirámide de la Luna: algunos resultados de la segunda temporada 1999', *Arqueología*, 23: 161–172.

Sugiyama, S. and Cabrera, R. (2006) 'El proyecto Pirámide de la Luna 1998-2004: conclusiones preliminares' in Sugiyama, S. and López Luján, L. (eds) *Sacrificios de consagración en la Pirámide de la Luna.* México: CONACULTA-Instituto Nacional de Antropología e Historia, Museo del Templo Mayor, Arizona State University: 11–24.

Sugiyama, S. and Cabrera, R. (2007) 'The Moon Pyramid Project and the Teotihuacan State Polity', *Ancient Mesoamerica*, 18(1): 109–125.

Sugiyama, S., Sugiyama, N. and Sarabia, A. (2014) 'El interior de la Pirámide del Sol en Teotihuacan', *Arqueología Mexicana*, 21(125): 24–29.

Torras Freixa, M. (2018) 'Rethinking monumentality in Teotihuacan, Mexico' in Brysbaert, A., Klinkenberg, V., Gutiérrez García-M, A. and Vikatou, I. (eds) *Constructing monuments, perceiving monumentality and the economics of building: theoretical and methodological approaches to the built environment.* Leiden: Sidestone Press: 219-242.

Torras Freixa, M. (2019) La construcción de una ciudad antigua en el Centro de México: planificación urbana y transformación social en Teotihuacan (1-250 d.C.). PhD Thesis. Barcelona: Universitat de Barcelona.

UNESCO (2012) *Inventaire de l'architecture de terre. Programme du patrimoine mondial pour l'architecture de terre (WHEAP).* CRATerre-ENSAG [online]. Available at: https://whc.unesco.org/fr/architecture-de-terre/ [accessed: 5 June 2019].